Microfluidics and Nanophotonics: Science and Engineering

Volume 4

This Springer book series focuses on the science and engineering of nanophotonics and miniaturized systems. It covers topics such as metamaterials, metasurfaces, biophotonics, microfluidics, Microelectromechanical systems (MEMS), and nano/microdevices. The series explores the translation of scientific breakthroughs in nanophotonics, metamaterials, and microfluidics into practical engineering applications, including metalens, micro-spectrometers, MEMS devices, biomedical devices and instrumentations, etc., to name a few. By featuring the latest research from leading international researchers and engineers, the series serves as a valuable resource for researchers, industry professionals, and undergraduate and graduate students interested in fundamental nanophotonics, cutting-edge micro/nanodevices, and related applications. More specifically, it includes the following themes:

- Metamaterials and Metasurfaces
- Nanoplasmonic Biosensing
- Biomedical Devices and Instrumentation
- Microsystems and MEMS
- Nanophotonics in Quantum Engineering

Sha Xiong · Ai-Qun Liu

Optofluidics: Fundamentals and Applications

Springer

Sha Xiong
School of Automation
Central South University
Changsha, Hunan, China

Ai-Qun Liu
Department of Electrical and Electronic Engineering
The Hong Kong Polytechnic University
Hong Kong, China

ISSN 2730-9029 ISSN 2730-9037 (electronic)
Microfluidics and Nanophotonics: Science and Engineering
ISBN 978-981-95-6295-4 ISBN 978-981-95-6296-1 (eBook)
https://doi.org/10.1007/978-981-95-6296-1

This Springer imprint is published by the registered company Springer Nature Singapore Pte Ltd.
The registered company address is: 152 Beach Road, #21-01/04 Gateway East, Singapore 189721, Singapore

Preface

Optofluidics integrates the fields of optics and microfluidics, providing novel functionalities with optical and fluidic property manipulations on the micro/nano meter scale. Studies on optofluidics have attracted engineers and scientists from various disciplines, such as chemistry, physics, materials and biology.

This book introduces the fundamentals and state-of-art for optofluidic technology and optofluidic devices, as well as its applications in biophysical analysis and detection. It covers the basic design principle and fabrication of optofluidic devices such as lens, grating and prism. It does not only demonstrate the optical property control with fluid manipulation, but also the optically driven fluids, such as laser induced jets, bubbles and nanodroplets in an optofluidic chip. The applications of optofluidic technology are highlighted in biophysical detection and analysis such as cell refractometer and bacteriophage detection.

The book is intended for undergraduate and graduate students who are interested in optofluidic technology, researchers investigating photonics and microsystems, and engineers working on lab on a chip or other microfluidic applications.

Changsha, China — Sha Xiong
Hong Kong, China — Ai-Qun Liu

Acknowledgments This book originates from the Ph.D. works of Dr. Sha Xiong, Dr. Lip Ket Chin, Dr. Jiaqing Yu, Dr. Zhenguo Li and Dr. Haitao Zhao, focusing on the fundamental researches and applications of optofluidic devices, which are led and fully supported by Prof. Ai-Qun Liu.

We would like to thank all the students, engineers, and colleagues from the VALENS center of Nanyang Technological University for their valuable assistance and help.

We would like to thank the staff from Springer, in particular, Dr. Mengchu Huang and Dr. Rammohan Krishnamurthy, for their help, support, and patience in the processing of manuscript submission.

We would like to thank the support from the National Natural Science Foundation of China (Grant No. 61905288), Natural Science Foundation of Hunan Province (Grant No. 2024JJ5447) and Natural Science Foundation of Changsha City (Grant No. kq2402228). Those research works are supported directly and indirectly by the Environmental and Water Industry Development Council of Singapore (Grant No. 1102-IRIS-05-01, 1102-IRIS-05-02, 1102-IRIS-05-04, MEWR C651/06/171), the AcRF Tier 1 grant from the Minister of Education of Singapore (Grant No. M4010891.120, RG45/09), the Science & Engineering Research Council of Singapore (Grant No. SERC 102 152 0018), National Research Foundation, Singapore, under the Competitive Research Program (Grant No. NRF2014NRF-CRP081-002).

Changsha, China — Sha Xiong
Hong Kong, China — Ai-Qun Liu

Competing Interests The authors have no competing interests to declare that are relevant to the content of this manuscript.

Contents

Chapter 1
Introduction of Optofluidics

1.1 Basic Concepts of Optofluidics

The term "optofluidics" first appeared at the start of this century to describe systems that integrate optics and fluidics. Initially, researchers defined optofluidics and categorized these systems based on their structure, mechanism, and functions [1–3]. Psaltis et al. identified three main types of interactions in optofluidics: structured solid–liquid hybrids where both media's optical properties are significant; fully fluid-based systems where only the fluids' optical properties matter; and colloid-based systems involving the manipulation of solid particles in liquid or utilizing the unique optical properties of colloidal solutions [1]. Monat et al. emphasized three primary applications of optofluidics: optofluidic light sources using liquid solutions as gain media; fluidic sensors with integrated microphotonic devices; and reconfigurable optical microdevices controlled by microfluids [2]. Even when the fluid does not play a central role in the device's operation or directly interact with light, indirect effects, such as thermally driven flow in microchannels [4], can still arise. Such interactions uphold the classification of the device as optofluidic, as the interplay between optical and fluidic components remains a defining characteristic.

In this book, optofluidics is described as a technology that manipulates light and fluids at the micro/nanoscale, leveraging their interaction to create versatile devices and integrated systems. We showcase optofluidic systems where fluids manipulate light or light manipulates fluids. Lab-on-a-chip technology, which utilizes microfluidic systems for biomedical and chemical analysis, has seen significant success [5, 6]. Integrating optical components into microfluidic systems enhances portability and sensitivity, overcoming the limitations of free-space optical elements. Typically, optofluidic devices involve adding optics to a microfluidic system or fluids to a photonic system [7] (Fig. 1.1). In some cases, the components are even made from liquids, providing remarkable flexibility and unique properties not found in solid equivalents.

S. Xiong and A.-Q. Liu, *Optofluidics: Fundamentals and Applications*,
Microfluidics and Nanophotonics: Science and Engineering 4,
https://doi.org/10.1007/978-981-95-6296-1_1

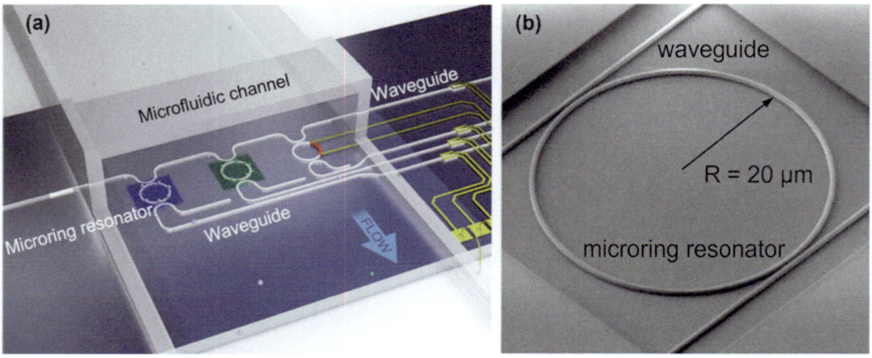

Fig. 1.1 **a** Schematic illustration of an optofluidic chip that consists of a microfluidic system and a microring resonator photonic system. **b** SEM image of the microring resonator structure on the chip. Reproduced from [7] with the permission of ACS publishing

Optofluidics has its roots in the development of microelectromechanical systems (MEMS), traditionally created using silicon-based micromachining processes. Early optofluidic devices were all based on MEMS fabrication techniques [8]. Over time, more affordable polymeric materials like polydimethylsiloxane (PDMS) and polymethylmethacrylate (PMMA) began to be used in microfluidic systems. PDMS is particularly preferred due to its optical transparency, biocompatibility, elasticity, and air permeability [9, 10]. PDMS chips are produced using soft lithography techniques, allowing for quick and inexpensive replication of identical devices. Glass is another favored material for microchips because of its chemical inertness, thermal stability, and physical robustness. The technology for fabricating glass nanochannels is well-established, offering high resolution and reproducibility. Today, glass is the most popular substrate for nanofluidic chip fabrication [11, 12]. Paper-based microfluidic device is a rising technique that uses patterned paper to split a single, low-volume (<40 μl) sample into multiple portions for parallel assays [13, 14]. These devices do not require external instrumentation or power, making them portable and disposable. Advances in fabrication technology have led to sophisticated optofluidic platforms with complex microfluidic networks. Optofluidic components can even be made from the same materials using the same processes, greatly enhancing system compatibility, integration, and efficiency.

Optofluidics has garnered significant research interest across various disciplines due to its unique advantages. The advanced microfluidic networks enable precise control for conducting multiple chemical analyses and biological diagnoses. These networks excel in handling fluids at the micron scale, allowing samples for optical analysis to be delivered, manipulated, and tested within a single compact chip. For instance, the optofluidic flow cytometer employs flow focusing technology to align cells and collects scattering and fluorescent signals as each cell passes through a laser beam [15]. The optical signals provide valuable information about both the physical properties (e.g., shape, size) and biochemical properties (e.g., protein content, cell cycle distribution) of cells, greatly enhancing the rapid and quantitative analysis of biological samples. Additionally, microfluidic networks enable parallel analysis.

Optofluidic systems offer unique advantages in creating reconfigurable optical components. The fluidic nature of liquids enables their use as tunable elements, while their smooth interfaces facilitate low-loss optical performance. By leveraging fluids with varying refractive indices, these systems can dynamically modulate optical properties. Examples of demonstrated optofluidic components include waveguides, microlenses, gratings, lasers, and interferometers.

Furthermore, optofluidic components often function as highly sensitive microsensors. For example, the optofluidic interferometer uses a fluid-air interface to create an optical path difference, suitable for biosensing or chemical analysis [16]. Other sensors include the optofluidic resonator, optofluidic microscope, and Bragg grating refractive-index sensors.

Lastly, optofluidic systems efficiently convert optical energy into mechanical, thermal, or chemical energy, aiding in hydrodynamic control, medical treatment, and biochemical analysis. Fluids can incorporate various optically active elements, such as gold nanoparticles, silica microspheres, and cells. Optical forces are used to manipulate and trap particles and target cells, and light can control flow by operating micropumps or generating cavitation bubbles. Optofluidic planar reactors uses TiO_2-coated micro-chamber for photocatalytic water treatment, effectively addressing the mass transfer and photon transfer limitations found in traditional photocatalytic reactors [17].

Additionally, optofluidic systems operate within a closed, miniaturized chip, allowing for the use of small sample volumes at low concentrations. This approach reduces costs, shortens running times, and provides a safer platform for biochemical studies. The enhanced portability and sensitivity of optofluidic devices are also anticipated to benefit environmental monitoring and clinical diagnostics.

A significant challenge in advancing optofluidics lies in enhancing control methods, which can be approached in two main ways. Firstly, fluids are used to manipulate light by creating versatile optical devices that are tunable, reconfigurable, and adaptive. There is an expectation to develop more types of optofluidic components with greater tunability. Secondly, light not only provides optical signals for analysis but also serves as an effective means to control fluid motion. Improving optical control, combined with the ability to manipulate small fluid volumes, could be a major step towards implementing optofluidic components in robust and highly automated applications.

1.2 Tunable Optofluidic Devices

Incorporating optical components on the microfluidic chips has proven to be an effective way to enhance the system's robustness, flexibility, and operator-independence. Additionally, this approach holds promise for the mass production of fully integrated systems. Fluid-configured optical components leverage three key advantages of microfluidics. Firstly, the flexible interfaces with sharp index contrast allow these devices to function similarly to their solid counterparts while offering high adaptability. Secondly, the index contrast can be modified by altering the fluid composition. The range of refractive index modulation achievable with microfluidics ($\Delta n \approx 0.5$–1) is significantly greater than that provided by electro-optic ($\Delta n \approx 10^{-3}$) or thermal effects ($\Delta n \approx 10^{-2}$) [2]. Thirdly, the mobility of fluids enables easy control of the interface shape and location at the microscale, making the devices reconfigurable.

The simplest application of laminar flow is the liquid–liquid waveguide [18] (Fig. 1.2a), where light is guided and confined within a liquid core that has a higher refractive index than the surrounding liquid cladding. Refractive-index gradients can be created by varying the concentration of solutions [19], dispersing particles in the core stream [20], or using thermal gradients [21]. Hydrodynamic control of liquid streams has been utilized to define the optical path. By reducing the flow rate ratio of the core to the cladding, the core's width decreases, thereby transforming the waveguide from multimode to single mode. An optical switch was demonstrated by directing the core stream into different outlets [22]. The liquid waveguide can also function as an optical splitter and a wavelength filter by controlling diffusion [23]. Furthermore, a transformation optics device was developed using an optofluidic waveguide, enabling chirped light focusing and strong interference effects with discrete diffraction [18]. Similarly, a Y-branch splitter (Fig. 1.2b) with large-angle bending and tunability was realized in an inhomogeneous medium through coordinate transformation [24].

Integrated microlenses are among the most extensively studied optical components. In-plane PDMS lenses are used to collimate light beams and enhance the incident light intensity for optical detection in adjacent microfluidic channels and chambers [25–27]. These lenses can be easily integrated onto a microfluidic chip, but their geometries are permanently fixed. Liquid lenses with adjustable focal lengths offer significant advantages in terms of functionality and versatility. Most out-of-plane lenses, which manipulate light paths perpendicular to the substrate plane, are designed as liquid-filled chambers covered by a thin membrane. The radius of curvature is controlled by the membrane's deflection when pressure is applied to the liquid [28]. Liquid lenses serve as a valuable complement to tunable microlenses driven by mechanisms such as microactuators or nonlinear effects, as they typically avoid the need for complex fabrication processes and precise alignment.

In-plane microlenses are particularly well-suited for lab-on-a-chip applications because they can be integrated with the microfluidic network on the same substrate. These lenses take advantage of the curvature formed at the interface between two immiscible fluids, such as liquid/liquid or liquid/gas. Dynamically reconfigurable liquid-core liquid-cladding lenses (Fig. 1.2c) have been developed using curved interfaces in rectangular chamber [29, 30] or circular chamber [31]. Liquid lens can also achieve the curved interface between two flow streams in a curved channel [32, 33]. Another type of liquid microlens was created using the liquid waveguide structure, operating at extremely low flow rates to achieve the desired refractive index gradient [34, 35]. With a transversely variable refractive index, a liquid gradient refractive index lens was formed. Light traveling along the optical axis of this gradient index

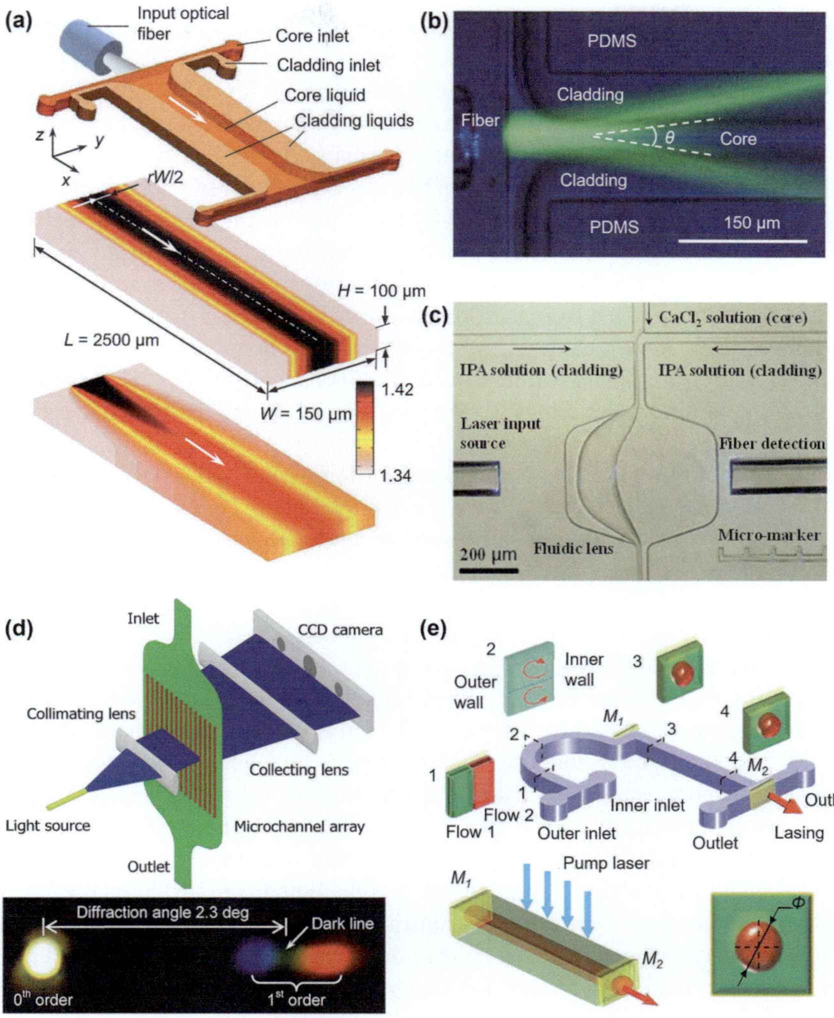

Fig. 1.2 **a** Schematic of an optofluidic waveguide with three laminar flows in the microchannel. Reproduced from [18] with permission from Springer Nature. **b** Photograph of the optofluidic Y-branch splitter. Reproduced from [24] with permission from the Royal Society of Chemistry. **c** Photograph of a liquid microlens. Reproduced from [30] with the permission of AIP publishing. **d** Schematic of a tunable transmission phase grating constructed by microchannel array with observed diffraction patterns. Reproduced from [36] with the permission of AIP publishing. **e** Schematic of the 3D liquid waveguide dye laser. Reproduced from [48] with the permission of the Royal Society of Chemistry

lens is gradually deviated toward the optical axis, resulting in a focused or collimated beam.

Optofluidic gratings are gaining significant attention for their use in filters and interferometers. These gratings can be composed of periodic structures involving liquid–solid, liquid–liquid, or liquid–gas combinations. One example is a transmission grating (Fig. 1.2d) created using an array of evenly spaced straight microchannels, where each grating line is formed by a flowing stream within a microchannel [36]. The diffraction pattern of the grating can be tuned by dynamically adjusting the refractive index of all the microflows simultaneously. Another example includes a diffraction grating made from uniform gas bubbles that self-assemble into ordered 2D lattices [37]. Additionally, 1D gratings have been demonstrated by aligning a stream of uniform droplets within a microfluidic channel [38–40]. The periodicity of these bubble or droplet arrays can be adjusted by varying the flow rate ratios of the two fluids.

In addition to basic optical components for tuning the optical path, integrating light sources into a microfluidic system is essential. Initially, micro-cavity liquid dye lasers were developed, with cavity sizes comparable to the wavelength of spontaneous emission [41, 42]. Subsequently, various configurations of optofluidic light sources have been explored. For instance, a mirroring-based laser demonstrates a switchable output with a highly multimode integrated waveguide, where the wavelength can be tuned by altering the dye/solvent properties [43]. Distributed-feedback dye lasers, which utilize higher-order Bragg gratings embedded in the liquid, produce narrow-band lasers by exploiting antiguiding effects and achieving single-mode operation [44, 45]. Another type of microfluidic light source uses liquid–liquid waveguides, formed by two immiscible liquid streams with different refractive indices [46]. The liquid core, filled with laser dye, is used for light emission and guiding. The laser beam size and intensity can be adjusted by varying the flow rate, while the emission wavelength is tuned by changing the composition of the liquid core. Additionally, broadband light sources can be created by constructing waveguide arrays and cascades [47]. A 3D liquid waveguide dye laser (Fig. 1.2e), which leverages centrifugal Dean flow to enhance light confinement, has been reported to be at least three times more efficient than its traditional 2D counterpart [48]. Dye lasers can also be implemented using droplet microfluidics [49]. Lasing occurs when the emitted light matches the whispering gallery mode of the droplet cavity. Fast-switching dye lasers utilize a stream of droplets containing different laser dyes, allowing the lasing wavelength to switch between 580 and 680 nm at frequencies up to 3.6 kHz [50]. Additionally, optofluidic ring resonator lasers based on micro-bubbles have been demonstrated [51]. These micro-bubbles, filled with a liquid gain medium, offer three-dimensional optical confinement, extremely high Q-factors, and the ability to handle liquids with different refractive indices.

With the advancement of various integrated optical elements, numerous optofluidic devices have been developed, including interferometers, optical switches, and attenuators, etc. These devices are crucial for incorporating optical detection and analytical techniques into microfluidic systems, which are used in biological and biomedical research. Examples include on-chip spectrometers, refractometers, and biomaterial concentration sensors. One notable example is the optofluidic microscope, which has successfully replaced bulky laboratory microscopes with a compact and portable imaging chip [52].

1.3 Optofluidic Applications

Tunable optofluidic components offer capabilities beyond simple light beam manipulation, finding extensive use in sensing and detection applications. This section examines the fundamental components (i.e., optofluidic waveguides and lenses), demonstrating how fluids can precisely control light propagation. Conversely, light can also be employed to manipulate fluid flows and suspended particles, enabling advanced optical sorting and targeting techniques. Furthermore, we explore the exciting application of cellular manipulation on optofluidic platforms.

Application of optofluidic waveguide

Optofluidic waveguides are essential components in integrated optofluidic systems. Incorporating fluidic elements not only enables light-matter interactions in areas that are challenging or impossible for solid waveguides, but also introduces additional fluidic-fluidic interactions. These features enhance the use of optofluidic waveguides in a wide range of applications, including on-chip optical devices, biochemical detection, and particle manipulation.

The primary function of optofluidic waveguides is to confine light energy within the waveguide core, a feature widely utilized for sample illumination. SU-8 waveguides are integrated into on-chip flow cytometers to excite target particles in the detection region [53]. Optofluidic waveguides can efficiently collect scattered light, increasing the signal-to-noise ratio by tenfold compared to conventional methods using a microscope objective [54]. They also work with beam splitters for multiplexed detection. With reconfigurable geometry and index profiles, optofluidic waveguides serve as fundamental components in many tunable optofluidic devices. The low-high-low index configuration of these waveguides is adapted for use as focusing lenses [55]. By adjusting the fluidic configuration, the focal distance can be varied two-dimensionally over several hundred micrometers. Additionally, optofluidic dye lasers based on 3D optofluidic waveguides have shown a threefold increase in slope efficiency compared to their traditional two-dimensional counterparts. Other integrated optofluidic components, such as optical switches [56], optical filters [57], and evanescent couplers [58], have also been developed.

Optofluidic waveguides serve as effective biochemical sensors owing to their enhanced light-matter interaction capabilities. These sensors employ diverse detection mechanisms. As illustrated in Fig. 1.3a, a liquid waveguide sensor is designed for concentration measurement [59]. Light is guided through the core layer while the analyte sample is introduced into one cladding layer, resulting in measurable alterations to the output light's modal power distribution that correlate with sample concentration. A common method involves detecting changes in the refractive index caused by target molecules. Optical signals vary with fluid properties based on light-matter interaction, even with minimal refractive index changes. Optofluidic waveguides often integrate with resonant structures to achieve high sensitivity. The detection limit in an antiresonant reflecting optical waveguide (ARROW) is 9×10^{-4} RIU by measuring resonance shifts in the transmission spectrum [60]. The limit improves to 1×10^{-7} RIU in a microsphere-based resonator [61]. Higher performance is possible with advanced resonators like whispering gallery mode and ring resonators. Another common approach is detecting the amount of absorbed light when it passes through the target solution, where absorption occurs if the optical frequency matches the molecular structures' resonant frequency. A bandgap hollow core waveguide, for example, can detect ethyl chloride at a concentration limit of 30 ppb [62]. Additionally, the light-focusing effect in optofluidic waveguides represents a distinctive optical phenomenon arising from gradual light refraction within a diffusion-induced gradient refractive index profile. Its dependence on molecular diffusion makes it a promising indicator for chemical processes. A characterization experiment confirms that focal length variations can effectively reflect diffusion properties [63]. As illustrated in Fig. 1.3b, this principle is applied to monitor sucrose hydrolysis reactions and determine diffusion coefficients of samples.

Optofluidic waveguides have a significant application in particle manipulation using the near-field evanescent wave. The electromagnetic field decays exponentially outside the waveguide core, creating a strong optical gradient force due to the intensity decay, making it ideal for nanoparticle manipulation. Solid SU-8 core and liquid cladding waveguides are used to trap and transport microparticles, achieving particle

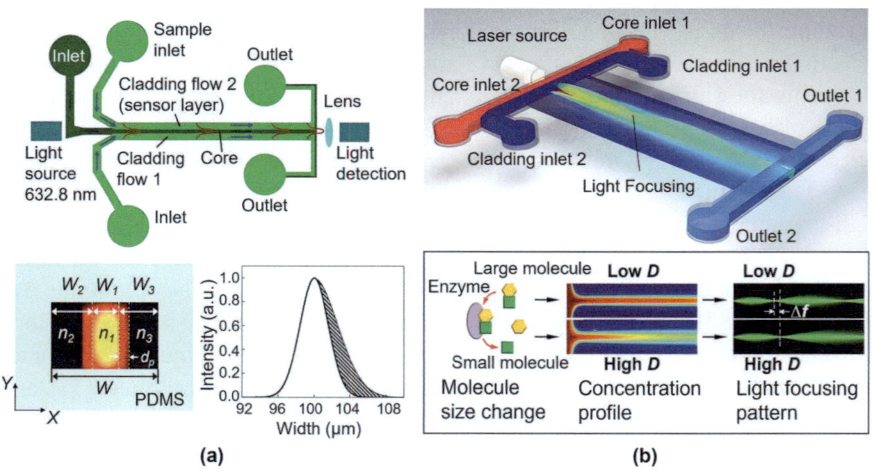

Fig. 1.3 **a** Illustration of the liquid/liquid waveguide sensor, including schematic top-view, cross-sectional view, and refractive index-dependent intensity profile variation. Reproduced from [59] with permission from AIP Publishing. **b** Optofluidic chip schematic demonstrating focal position modulation corresponding to average diffusion coefficient changes during chemical reaction progression. Reproduced from [63] with permission from Elsevier

velocities of 28 μm/s with a guided power of 53.5 mW [64]. Nanoparticle trapping is demonstrated in waveguides with intensity enhancement structures, successfully trapping 75-nm nanoparticles and λ-DNA molecules are trapped in sub-wavelength slot waveguides [65]. This trapping occurs along a line, allowing for direct handling of extended biomolecules. A key advantage of optofluidic waveguide-based particle manipulation is that all molecules are trapped on an image plane, facilitating experimental observation and quantitative analysis. Additionally, parallel DNA molecule trapping is achieved in a controlled manner using the standing-wave evanescent field [66]. Moreover, optofluidic waveguides can be combined with various detection techniques, enhancing their potential for biomedical applications. However, a significant challenge with waveguide-based particle manipulation is that the effective manipulation region is confined to 100–400 nm above the waveguide surface. To improve efficiency, shallow microchannels are required to bring target molecules into the optical interaction volume, which complicates the fabrication process and flow control. Another challenge is the presence of complex hydrophobic and electrostatic interactions at the solid–liquid interface. Therefore, careful preparation of the buffer solution and surface conditions is necessary to achieve a noticeable optical force effect.

Application of optofluidic lens

Optofluidic lenses offer several advantages over solid lenses, including high flexibility, compact size, and compatibility with microfluidic systems. They eliminate the need for assembly and alignment required in solid lenses. These lenses meet the increasing demand for lab-on-a-chip systems and are utilized in various applications such as imaging, sample illumination, optical detection, and particle manipulation.

Optical imaging is a key function for both solid and optofluidic lenses, requiring high performance in terms of stability, fast response time, and minimal optical aberrations. A tunable lens array has been demonstrated using soft lithographic techniques, where a PDMS membrane is bonded to an underlying microfluidic network [67]. By controlling external pressure, the imaging plane can be adjusted over several hundred micrometers, achieving an optical resolution of 3 μm at 20 kPa. Unlike solid optical systems, optofluidic lenses cannot correct optical aberrations using additional compensation lenses. Therefore, to enhance imaging quality, optical aberrations must be deliberately minimized through other means. An optofluidic lens with a tunable focal length and asphericity has been shown to reduce spherical aberration [68]. This is achieved by controlling the interface between a conductive fluid and a non-conductive ambient fluid through variations in hydrostatic pressure and electric fields.

Optofluidic lenses are also utilized for sample illumination in optical detection. On-chip flow cytometers incorporate liquid–liquid lenses to minimize light beam divergence or scattering [69]. In an optofluidic Fabry–Pérot (FP) resonator, a pair of liquid microlenses are employed to achieve efficient light coupling between an optical fiber and the microcavity (Fig. 1.4a). These microlenses collimate the diverging beam, enabling real-time optimization of the cavity finesse, which can reach up to

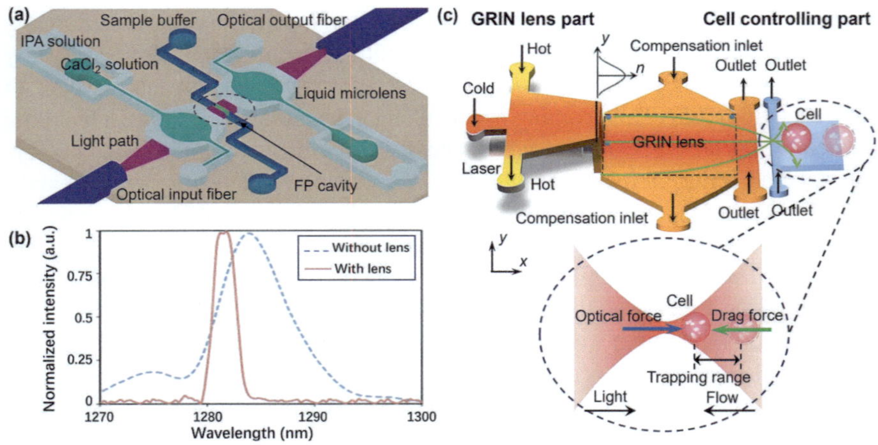

Fig. 1.4 **a** Schematic illustration of the optofluidic FP resonator. **b** The transmission spectra of the optofluidic FP resonator without microlenses and with microlenses. Reproduced from [70] with permission from AIP Publishing. **c** Schematic illustration of the liquid thermal GRIN lens for single cell trapping. Reproduced from [72] with permission from the Royal Society of Chemistry

18.79 [70]. As demonstrated in Fig. 1.4b, the incorporation of microlenses significantly enhances the quality factor of the FP cavity, thereby improving detection sensitivity. A gradient-index lens exhibiting minimized spherical aberration and field curvature has been experimentally validated to significantly enhance focusing performance [55]. Through precise optical design, this lens achieves a reduced beam waist diameter of 7.8 μm, thereby optimizing key analytical parameters in flow cytometry systems, including throughput enhancement, improved signal-to-noise ratio (SNR), and coefficient of variation (CV) reduction. Beyond flow cytometry applications, the lens demonstrates versatile utility in high-precision measurements, such as particle velocity tracking, multiplexed barcoded particle scanning, and quantitative photobleaching analysis.

The optofluidic lens serves a dual function in microfluidic systems: it enables both sample excitation and simultaneous collection of fluorescent/scattering signals from the target detection region. A droplet-based lens array integrated into a microfluidic device enhances fluorescent intensity, increasing fluorescent and spatial resolution by eightfold compared to devices without the lens array [71]. The fluidic lens in on-chip flow cytometry enables angularly resolved light scattering detection by selectively capturing signals from predefined angular regions, thereby facilitating high-resolution particle characterization within microfluidic channels [25]. This integrated optical architecture eliminates the need for separate excitation and detection components, thereby streamlining system complexity.

Another significant application of optofluidic lenses is their integration with optical tweezers. A tunable thermal GRIN lens has been demonstrated to trap single living cells in flowing environments (Fig. 1.4c), with a focal length adjustable from 430 to 930 μm and a beam waist of 4 μm [72]. This allows for rapid and straightforward cell manipulation by adjusting the focal length.

Optofluidic lens can also serve as sensors for biomedical detection. Their focal length is influenced by external factors such as pH, temperature, and protein levels. For example, an optical pressure sensor operates on the principle that a variation in pressure alters the curvature of liquid lens, thereby affecting its focal length [73]. This change can be measured by capturing and analyzing interferometric images. Similarly, the thermal lens effect is utilized in microscopy by analyzing signal and phase dependencies on optical configurations between the sample and the probe beam focus, enabling detection of molecular distribution or individual nanoparticles [74].

Optical manipulation

Besides illumination and excitation, light is utilized for particle manipulation on optofluidic devices. Recently, optical forces have gained attention as a non-contact method for particle trapping and separation. These forces are primarily divided into propagation force and gradient force. When photons are absorbed or scattered by particles, they transfer momentum, causing the propagation force to move particles in the direction of light propagation. Conversely, particles are drawn along the gradient to the area with the strongest electric field, with the gradient force directing them to the region of highest optical intensity.

A tightly-focused single beam is designed for submicron particles [75] and is used to investigate weak signals from individual particles or cells, such as scattering spectra [76] and Raman spectra [77], due to improved noise reduction. Dual-beam setups have shown significant advancements in cell manipulation [78, 79]. The divergent fields can create a large area trap for relatively large targets (e.g., cells), providing greater positional control and reducing the risk of damage from high-energy lasers. Beam arrays offer greater flexibility in tuning potential barriers and interconnecting optical traps for particle sorting [80, 81]. Combining weakly focused beams with opposing fluid flows in microfluidic systems has demonstrated particle separation and sorting [82–84]. The optofluidic lattice provides deterministic control over particle dynamics, including aggregation modulation and multi-hopping phenomena [85]. This platform further facilitates ultra-sensitive biosensing by enabling single-bacteria screening of biomolecular interactions (e.g., antibody-antigen binding) through precisely particle-enabled bacteria hopping, as illustrated in Fig. 1.5a.

When a waveguide is combined with optical tweezers, evanescent fields have been used to guide and trap particles and cells [64, 86, 87]. This approach offers convenient manipulation, improved stability, and greater spatial localization through the near field, although the working zone is limited to a few hundred nanometers from the surface. The slot waveguides successfully captured polystyrene particles and RNA particles tens of nanometers in diameter [65]. Precisely engineered adjacent waveguides generate a dense array of subwavelength optical hotspots (potential wells) with uniform field distributions. Through dynamic modulation of incident laser intensity, the dual waveguide array architecture achieves nanoparticle sorting

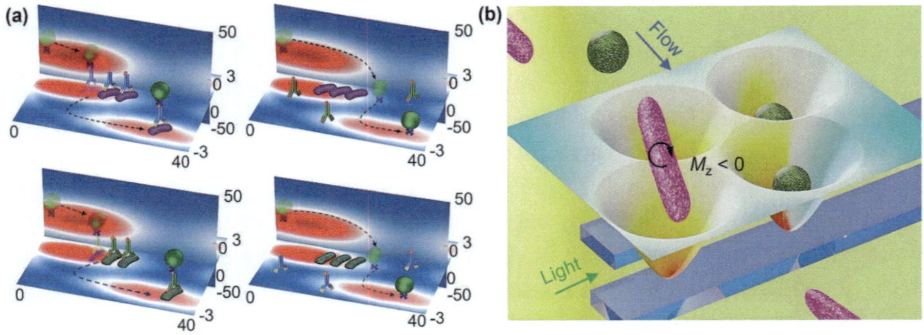

Fig. 1.5 **a** Screening for bacterial antibodies with optofluidic tweezers with specific antibodies. Reproduced from [85] with permission from Springer Nature. **b** Waveguide optical tweezers capture bacteria by shape. Reproduced from [89] with permission from American Chemical Society

(100–500 nm) by exploiting the energy gradient between adjacent optical potential wells [88]. A photonic nanowaveguide array manipulates coupled optical hotspots to exert shape-dependent torques on submicron particles (Fig. 1.5b). Spherical bacteria (e.g., *S. aureus*) are confined to individual potential wells due to their size matching the well dimensions, while rod-shaped bacteria (e.g., *E. coli*) experience torque from adjacent wells, enabling their separation despite similar volumes [89].

Beyond directly manipulating particles with optical forces, lasers can also control fluids by operating pumps and mixers. Colloidal particles have been assembled in microfluidic channels and manipulated using optical forces to function as pumps [90, 91]. Additionally, the rotation of microgears [92] or rotors [93] using optical forces has been demonstrated. Optically actuated pumps offer benefits such as easy reconfigurability, seamless integration and compact size compared to other micropumps.

The photothermal effect is a fascinating phenomenon that converts optical energy into hydrodynamic energy for optofluidic applications [94, 95]. By using focused laser beams with photothermal nanoparticles (such as gold nanoparticles), liquid flows can be driven and directed in microfluidic channels to transport biomolecules and cells at controlled speeds and directions. When a laser illuminates a liquid with photothermal particles near the liquid–air interface, it generates heat, accelerating evaporation and producing vapor. The vapor condenses into droplets, which merge back into the liquid, causing the contact line to expand. It has been observed that the optofluidic flow speed in a channel 10 μm wide can reach up to 500 μm/s [4].

Light scattering by variations in a fluid's refractive index can drive large-scale flow [96]. When a laser beam passes through the liquid, it deforms the interface, creating a liquid jet. This scattering generates an upward force, causing an upward flow in the illuminated area. This flow is then refilled by a downward flow, forming a toroidal recirculation, which deforms the interface due to viscous stresses.

Laser-induced cavitation is another way to achieve optical-hydrodynamic energy conversation. Due to the rapid expansion and collapse of cavitation bubbles, along with their flow-focusing capabilities, liquids can be accelerated to extremely high speeds. In microfluidic channels, laser pulses create cavitation bubbles that disrupt the laminar flow of two fluids, forming a localized mixing zone that enhances the rapid mixing of chemicals in lab-on-a-chip devices [97]. A micropump driven by laser-induced bubbles has been shown to offer consistent and precise control over picolitre volumes [98]. The transient radial flow produced by these bubbles is also employed to manipulate microscale particles [99] and nanowires [100]. Furthermore, cavitation aids in various cell operations, such as cell surgery and lysis [101, 102].

Cell on chips

Numerous integrated optofluidic devices have been shown to handle and analyze cells effectively. Microchips serve as platforms for various cell-related processes, including cell culture, treatment, selection, lysis, and analysis [103].

In vitro cell culture is a crucial field in modern biology, but challenges in mimicking in vivo environmental conditions have limited research progress. Microfluidic systems provide a viable solution with improved environmental controls. On-chip tissue organization and cell culture can be achieved by integrating cell delivery mechanisms with on-chip cell seeding, scaffolds, or extracellular matrices for cell growth, as well as systems for injecting culture media and soluble factors for cell stimulation [104, 105]. Microfabrication techniques, such as soft-lithography, photolithography, and e-beam patterning, enable the creation of nano- to micro-sized surface patterns. These features confine and control cell ensembles and their interactions with the extracellular matrix. Additionally, surface chemistry modifications enhance and regulate cell attachment within defined geometrical patterns [106]. Precise control of the cell seeding environment allows for studies on how cell growth behaviors are influenced by the shape and size of cell ensembles. The microfluidic system provides precise spatial and temporal control over culture media and soluble factors, allowing systematic investigation of cellular phenotypes and behaviors through controlled extracellular perturbations [107, 108]. As shown in Fig. 1.6, a microfluidic network with an array of cell seeding sites enables multi-condition testing, while a pulsation-free pump regulates flow rate and medium frequency to replicate physiological blood flow profiles [109]. Integrated biosensors continuously monitor critical parameters, including pH, calcium ion concentration, dissolved oxygen, and biomass.

Cultured cells can be fractionated or selected using optofluidic technologies to analyze specific cell types separately. Adherent cell groups can be treated with Trypsin solution to detach and harvest the cells [110]. The suspended cells are then directed through a cell sorting mechanism to isolate the desired cell type. Various sorting mechanisms have been developed for microfluidic chips, including dielectrophoretic (DEP) cell sorters, fluorescently activated cell sorters (FACS), and surface affinity-based trapping techniques [111–113]. FACS utilizes the fluorescent properties of tagged particles to identify and divert the target cell type into a collection chamber using integrated pumps and valves. These sorters are commonly used in flow cytometers, and microfluidic systems can miniaturize these devices into portable formats. Additionally, various optofluidic tweezers are widely used for cell trapping and sorting. Optical lattices enable continuous particle fractionation based on size

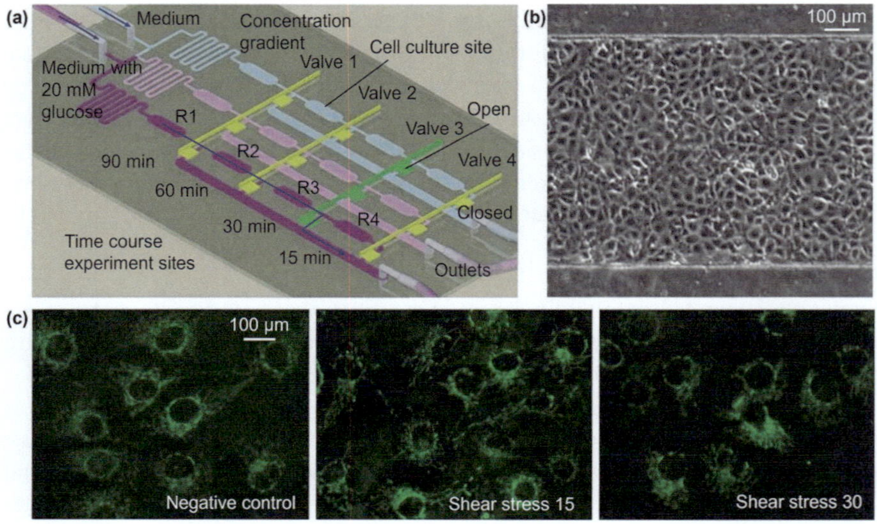

Fig. 1.6 **a** Schematic illustrations of the microfluidic chip with configurable flow profile to mimic the blood flow in a blood vessel. **b** Images of endothelial cells cultured in the microchannel for 48 h. **c** Mitochondrial morphology of endothelial cells under different shear treatment conditions. Reproduced from [109] with permission from the Royal Society of Chemistry

and refractive index with high throughput [114]. When mixed particles flow through the lattice, the target ones deviate and follow the designed trajectories. Optical chromatography exploits the synergistic interaction of optical scattering force and fluidic forces to achieve cell separation with high resolution [82, 84].

Cell lysis is a widely used technique to break the cell membrane without denaturing or degrading sensitive macromolecules, such as DNA and proteins. This process can be achieved through enzymatic, osmotic, chemical, or mechanical methods. Enzymatic lysis, such as using lysozyme, is commonly employed to break down bacterial cells [115]. Cytolysis occurs when cells are placed in a hypotonic environment, causing the cell membrane to burst due to osmotic pressure [116]. Chemical lysis can be performed using agents like Triton X-100 and sodium dodecyl sulfate in micro-environments [117, 118]. Mechanical lysis involves applying shear stress to rupture the cell membrane [119]. Laser-induced cavitation bubbles can effectively rupture the cell membrane and create single-cell poration [120]. The released proteins and DNA from lysed cells can then be analyzed.

For cell-based optofluidic devices, various biosensors have been developed for applications such as drug screening, single-cell analysis, and cell–cell interaction. These biosensors typically generate readouts based on electrical or optical signals. Electrical signals, which arise from the cell's chemical-dependent electrophysiological activity, are monitored and recorded using microfluidic devices integrated with microelectrode arrays. For optical approaches, image-based techniques and intracellular fluorescent probes are commonly used for single-cell analysis. Researchers are extensively studying the integration of optical elements into microfluidic devices to enable optical manipulation and detection, thereby reducing reliance on free-space optical elements or microscopic systems.

1.4 Future

Recent advancements in fundamental optofluidic components have led to the development of various tools for biochemical analysis and treatment, featuring miniaturized and simplified designs. These components are highly reconfigurable, offering significant tunability and making the optical system extremely adaptable. This adaptability, coupled with miniaturization, can greatly reduce the overall system size, making it portable for field testing. By integrating optics into the system, rather than relying on bulky free-space optics, costs can be significantly lowered, paving the way for affordable point-of-care diagnostics, even in low-resource settings. It is expected that a fully integrated optofluidic system, with all optical functions on a single chip, will be achieved in the near future. These systems hold great potential for enhancing test speed, accuracy, and throughput.

Optofluidic devices have progressed to the stage where they can accurately and selectively detect a wide range of analytes in various biofluids such as blood, saliva, and urine. The initial steps towards measuring multiple analytes in complex samples have already been taken. The next step is for these systems to perform multi-target assays, enabling the simultaneous detection of different viruses, ions, and diverse entities such as viral DNA, proteins, and mRNA within a single assay. These multi-target assays are essential for diagnosing diseases like cancer and infections, as accurate diagnosis often requires detecting multiple biomarkers rather than just one. While cell-based biomarkers have greatly advanced disease diagnosis, their clinical application is hindered by complex cell handling procedures and the need for large-scale microscopy systems in functional assays. We foresee that optofluidic systems could offer a promising solution to transform cell functional assays into practical tools for disease diagnosis.

Water analysis employs well-established methods for detecting diverse compound classes, yet these typically require multiple analyses to quantify individual compounds. Conventional techniques, such as high-resolution liquid chromatography and mass spectrometry, are time-consuming and costly. Moreover, they require specialists or technicians to simultaneously analyze and separate a large number of compounds. Optofluidic systems aim to provide portable instruments for on-site sampling and real-time automatic online analysis.

Beyond water and biofluid analysis, optofluidic sensors have demonstrated utility in food matrices, e.g., identifying adulterants in beverages like soft drinks or milk. Moreover, these devices show promise for aerosol and gas sensing, presenting compelling avenues for future investigation. Efforts toward these directions will help unlock the full potential of optofluidic technologies for biochemical applications and support their widespread adoption in clinical and industrial practices.

Identification and characterization of single bio-objects is an attractive research direction that often requires the manipulation of individual micro/nanoparticle. While laser beam trapping is highly effective for single cells, manipulating and detecting single biomolecules is challenging due to their nanoscale size, which exceeds the diffraction limit. By integrating nanostructures like nanowaveguides, and nanoholes into devices, nanoplasmonic and photonic crystal resonator-based biosensors can manipulate and detect nano-objects, including biomolecules. However, the potential of advanced nanophotonic devices, such as metasurface-based biosensing, as analytical platforms for optofluidic applications has yet to be fully explored.

Unlike conventional glass-based optical elements that manipulate light in free space, achieving high-numerical-aperture beam focusing with optofluidic components presents significant challenges. This limitation may hinder the development of high-strength and high-resolution optical manipulation and detection. To address this, novel fabrication methods or materials for on-chip light control need to be explored, such as waveguide using femtosecond laser direct writing techniques. Additionally, improvements in the optical performance of optofluidic components, including the bandwidth of on-chip photodetectors and light coupling efficiency into waveguides, are necessary. These advancements will require progress in material science, optical engineering, and micro/nanomanufacturing.

Despite advancements, most optofluidic devices still depend on external and complex instrumentation. This reliance makes them more of a specialized analytical tool rather than a portable, stand-alone system suitable for point-of-care applications. Many existing optofluidic systems need various external instruments for flow control and signal detection, posing a significant challenge in developing all-in-one systems capable of performing the entire assay. In the near future, integrating all optical, electronic, and flow control components into a portable system, along with automated signal readout and analysis, could be a promising solution. Furthermore, with the development of optofluidic technology, integrating cost-effective optical components, such as organic photodiodes and reproducible lasers, and combining pump-free microfluidic technology onto a disposable microchip for rapid, high-sensitivity, and low-cost assays, will bring a bright future for optofluidic applications. Achieving this will require collaborative efforts from both academia and industry.

The integration of artificial intelligence (AI) and optofluidics is catalyzing transformative breakthroughs in biomedical, chemical, and environmental fields. By using machine learning (ML) algorithms, AI optimizes microfluidic chip designs and processes vast, multidimensional datasets to reveal hidden patterns, enabling real-time adaptive control for experiments. This synergy powers intelligent platforms for chemical synthesis, reaction optimization, and biological assay design. Deep learning (DL) models further enhance real-time analysis of light-matter interactions in optofluidic systems, facilitating applications like high-throughput live-cell 3D imaging and droplet scattering pattern prediction in microchannels. While challenges remain in data quality and computational efficiency, the rapid pace of innovation suggests these barriers will be overcome. The next decade will likely witness widespread adoption across research, industrial, and clinical applications, fulfilling the vision of smart lab-on-a-chip systems that accelerate scientific discovery and improve human health.

1.5 A Overview of Book Contents

This book focuses on the pioneer works based on fundamental optofluidic devices and their potential applications, organized as follows.

Chapter 1 Introduction of Optofluidics. This chapter briefly introduces basic concepts of optofluidics, tunable optofluidic devices with examples, and optofluidic applications. The future trends and challenges of optofluidics are discussed.

Chapter 2 Optofluidic Microlens. This chapter focuses on optofluidic microlens based on co-flow focusing. The discussion covers the working principle, the fabrication processes, focusing power of the microlens, and its effectiveness for light collimation.

Chapter 3 Optofluidic Prism. This chapter focuses on optofluidic prism realized by laminar flow streams. The discussion covers the working principle, shape control of the prism and corresponding changes of the deviation angle of the light beam, light scanning, and refractometer based on the optofluidic prism.

Chapter 4 Optofluidic Hyperbolic Secant Lens. This chapter focuses on optofluidic hyperbolic secant lens based on gradient refractive index (GRIN). The discussion covers the working principle, key properties (i.e., optical aberrations, index profile) and performance evaluation of optofluidic hyperbolic secant lens.

Chapter 5 Optofluidic Droplet Grating. This chapter focuses on optofluidic grating including diffraction grating and long-period grating using microfluidic droplets. The discussion covers the working principle, droplet formation, relationship between the optical signals (diffraction or transmission) and the droplet grating characteristics.

Chapter 6 Optofluidic Nanobubble and Nanodroplet. This chapter focuses on nanobubble and nanodroplet generated by laser on an optofluidic chip. The discussion covers the working principle, bubble dynamics in confined channel, jets from bubble interaction, breakup of nano-thickness film and nanodroplet generation process.

Chapter 7 Optofluidic Droplet Fusion. This chapter focuses on droplet fusion by creating a cavitation bubble on an optofluidic chip. The discussion covers the working principle, droplet fusion under different conditions, and fusion of droplets with bubble or cell.

Chapter 8 Optofluidic Measurement with Imaging Techniques. This chapter focuses on optofluidic measurement methods with the facility of imaging techniques. The discussion covers the refractive index measurement and tensile strength measurement. On each part, the working principle and case study are provided.

Chapter 9 Optofluidic Devices. This chapter focuses on the optofluidic devices based on the integration of varies modules and multifunction. The discussion covers the working principle and experimental results of optofluidic devices for refractive index measurement, cell poration, and virus detection.

References

1. Psaltis D, Quake SR, Yang C (2006) Developing optofluidic technology through the fusion of microfluidics and optics. Nature 442:381–386
2. Monat C, Domachuk P, Eggleton BJ (2007) Integrated optofluidics: a new river of light. Nat Photon 1(2):106–114
3. Horowitz VR, Awschalom DD, Pennathur S (2008) Optofluidics: field or technique? Lab Chip 8(11):1856–1863
4. Liu GL, Kim J, Lu Y, Lee LP (2006) Optofluidic control using photother-mal nanoparticles. Nat Mater 5(1):27–32
5. Stone HA, Stroock AD, Ajdari A (2004) Engineering flows in small devices: microfluidics toward a lab-on-a-chip. Annu Rev Fluid Mech 36:381–411
6. Whitesides GM (2006) The origins and the future of microfluidics. Nature 442:368–373
7. Feng X, Zhang G, Chin LK, Liu AQ, Liedberg B (2017) Highly sensitive, label-free detection of 2,4-dichlorophenoxyacetic acid using an optofluidic chip. ACS Sens 2:955–960
8. Grayson ACR, Shawgo RS, Johnson AM, Flynn NT, Li YW, Cima MJ, Langer R (2004) A BioMEMS review: MEMS technology for physiologically integrated devices. Proc IEEE 92(1):6–21
9. Ng JMK, Gitlin I, Stroock AD, Whitesides GM (2002) Components for integrated poly(dimethylsiloxane) microfluidic systems. Electrophoresis 23(20):3461–3473
10. Whitesides GM, Stroock AD (2001) Flexible methods for microfluidics. Phys Today 54(6):42–48
11. Mawatari K, Kubota S, Xu Y, Priest C, Sedev R, Ralston J, Kitamori T (2012) Femtoliter droplet handling in nanofluidic channels: a Laplace nanovalve. Anal Chem 84(24):10812–10816
12. Xu Y, Wang CX, Li LX, Matsumoto N, Jang K, Dong YY, Mawatari K, Suga T, Kitamori T (2013) Bonding of glass nanofluidic chips at room temperature by a one-step surface activation using an O-2/CF4 plasma treatment dagger. Lab Chip 13(6):1048–1052
13. Pollock NR, Rolland JP, Kumar S, Beattie PD, Jain S, Noubary F, Wong VL, Pohlmann RA, Ryan US, Whitesides GM (2012) A paper-based multiplexed transaminase test for low-cost, point-of-care liver function testing. Sci Transl Med 4(152):152ra129
14. Mazzeo AD, Kalb WB, Chan L, Killian MG, Bloch JF, Mazzeo BA, Whitesides GM (2012) Paper-based, capacitive touch pads. Adv Mater 24(21):2850–2856
15. Cho SH, Godin JM, Chen CH, Qiao W, Lee H, Lo YH (2010) Review article: recent advancements in optofluidic flow cytometer. Biomicrofluidics 4(4):43001
16. Grillet C, Domachuk P, Ta'eed V, Magi E, Bolger JA, Eggleton BJ, Rodd LE, Cooper-White J (2004) Compact tunable microfluidic interferometer. Opt Express 12(22):5440–5447
17. Lei L, Wang N, Zhang XM, Tai QD, Tsai DP, Chan HLW (2010) Optofluidic planar reactors for photocatalytic water treatment using solar energy. Biomicrofluidics 4(4):043004
18. Yang Y, Liu AQ, Chin LK, Zhang XM, Tsai DP, Lin CL, Lu C, Wang GP, Zheludev NI (2012) Optofluidic waveguide as a transformation optics device for lightwave bending and manipulation. Nat Commun 3:651
19. Brown M, Vestad T, Oakey J, Marr DWM (2006) Optical waveguides via viscosity-mismatched microfluidic flows. Appl Phys Lett 88(13):134109
20. Conroy RS, Mayers BT, Vezenov DV, Wolfe DB, Prentiss MG, Whitesides GM (2005) Optical waveguiding in suspensions of dielectric particles. Appl Opt 44(36):7853–7857
21. Tang SKY, Mayers BT, Vezenov DV, Whitesides GM (2006) Optical waveguiding using thermal gradients across homogeneous liquids in microfluidic channels. Appl Phys Lett 88(6):061112
22. Wolfe DB, Conroy RS, Garstecki P, Mayers BT, Fischbach MA, Paul KE, Prentiss M, Whitesides GM (2004) Dynamic control of liquid-core/liquid-cladding optical waveguides. Proc Natl Acad Sci USA 101(34):12434–12438
23. Wolfe DB, Vezenov DV, Mayers BT, Whitesides GM, Conroy RS, Prentiss MG (2005) Diffusion-controlled optical elements for optofluidics. Appl Phys Lett 87(18):181105

24. Yang Y, Chin LK, Tsai JM, Tsai DP, Zheludev N, Liu AQ (2012) Transformation optofluidics for large-angle light bending and tuning. Lab Chip 12(19):3785–3790
25. Godin J, Lien V, Lo YH (2006) Demonstration of two-dimensional fluidic lens for integration into microfluidic flow cytometers. Appl Phys Lett 89(6):061106
26. Camou S, Fujita H, Fujii T (2003) PDMS 2D optical lens integrated with microfluidic channels: principle and characterization. Lab Chip 3(1):40–45
27. Wang Z, El-Ali J, Engelund M, Gotsaed T, Perch-Nielsen IR, Mogensen KB, Snakenborg D, Kutter JP, Wolff A (2004) Measurements of scattered light on a microchip flow cytometer with integrated polymer based optical elements. Lab Chip 4(4):372–377
28. Pang L, Levy U, Campbell K, Groisman A, Fainman Y (2005) Set of two orthogonal adaptive cylindrical lenses in a monolith elastomer device. Opt Express 13(22):9003–9013
29. Tang SKY, Stan CA, Whitesides GM (2008) Dynamically reconfigurable liquid-core liquid-cladding lens in a microfluidic channel. Lab Chip 8(3):395–401
30. Seow YC, Liu AQ, Chin LK, Li XC, Huang HJ, Cheng TH, Zhou XQ (2008) Different curvatures of tunable liquid microlens via the control of laminar flow rate. Appl Phys Lett 93(8):084101
31. Song C, Nguyen NT, Tan SH, Asundi AK (2009) Modelling and optimization of micro optofluidic lenses. Lab Chip 9(9):1178–1184
32. Mao XL, Waldeisen JR, Juluri BK, Huang TJ (2007) Hydrodynamically tunable optofluidic cylindrical microlens. Lab Chip 7(10):1303–1308
33. Rosenauer M, Vellekoop MJ (2009) 3D fluidic lens shaping—a multiconvex hydrodynamically adjustable optofluidic microlens. Lab Chip 9(8):1040–1042
34. Huang H, Mao XL, Lin SCS, Kiraly B, Huang YP, Huang TJ (2010) Tunable two-dimensional liquid gradient refractive index (L-GRIN) lens for variable light focusing. Lab Chip 10(18):2387–2393
35. Mao XL, Lin SCS, Lapsley MI, Shi JJ, Juluri BK, Huang TJ (2009) Tunable Liquid Gradient Refractive Index (L-GRIN) lens with two degrees of freedom. Lab Chip 9(14):2050–2058
36. Li ZG, Yang Y, Zhang XM, Liu AQ, Zhang JB, Cheng L, Li ZH (2010) Tunable visual color filter using microfluidic grating. Biomicrofluidics 4(4):043013
37. Hashimoto M, Mayers B, Garstecki P, Whitesides GM (2006) Flowing lattices of bubbles as tunable, self-assembled diffraction gratings. Small 2(11):1292–1298
38. Yu JQ, Yang Y, Liu AQ, Chin LK, Zhang XM (2010) Microfluidic droplet grating for reconfigurable optical diffraction. Opt Lett 35(11):1890–1892
39. Chin LK, Liu AQ, Zhang JB, Lim CS, Soh YC (2008) An on-chip liquid tunable grating using multiphase droplet microfluidics. Appl Phys Lett 93(16):164107
40. Chin LK, Liu AQ, Soh YC, Lim CS, Lin CL (2010) A reconfigurable optofluidic Michelson interferometer using tunable droplet grating. Lab Chip 10(8):1072–1078
41. Helbo B, Kristensen A, Menon A (2003) A micro-cavity fluidic dye laser. J Micromech Microeng 13(2):307–311
42. Galas JC, Torres J, Belotti M, Kou Q, Chen Y (2005) Microfluidic tunable dye laser with integrated mixer and ring resonator. Appl Phys Lett 86(26):264101
43. Gersborg-Hansen M, Balslev S, Mortensen NA, Kristensen A (2005) A coupled cavity microfluidic dye ring laser. Microelectron Eng 78–79:185–189
44. Balslev S, Kristensen A (2005) Microfluidic single-mode laser using high-order Bragg grating and antiguiding segments. Opt Express 13(1):344–351
45. Li ZY, Zhang ZY, Scherer A, Psaltis D (2006) Mechanically tunable optofluidic distributed feedback dye laser. Opt Express 14(22):10494–10499
46. Vezenov DV, Mayers BT, Wolfe DB, Whitesides GM (2005) Integrated fluorescent light source for optofluidic applications. Appl Phys Lett 86(4):041104
47. Mayers BT, Vezenov DV, Vullev VI, Whitesides GM (2005) Arrays and cascades of fluorescent liquid-liquid waveguides: broadband light sources for spectroscopy in microchannels. Anal Chem 77(5):1310–1316
48. Yang Y, Liu AQ, Lei L, Chin LK, Ohl CD, Wang QJ, Yoon HS (2011) A tunable 3D optofluidic waveguide dye laser via two centrifugal Dean flow streams. Lab Chip 11(18):3182–3187

49. Tang SKY, Derda R, Quan QM, Loncar M, Whitesides GM (2011) Continuously tunable microdroplet-laser in a microfluidic channel. Opt Express 19(3):2204–2215
50. Tang SKY, Li ZY, Abate AR, Agresti JJ, Weitz DA, Psaltis D, Whitesides GM (2009) A multi-color fast-switching microfluidic droplet dye laser. Lab Chip 9(19):2767–2771
51. Lee W, Sun YZ, Li H, Reddy K, Sumetsky M, Fan XD (2011) A quasi-droplet optofluidic ring resonator laser using a micro-bubble. Appl Phys Lett 99(9):091102
52. Heng X, Erickson D, Baugh LR, Yaqoob Z, Sternberg PW, Psaltis D, Yang CH (2006) Optofluidic microscopy—a method for implementing a high resolution optical microscope on a chip. Lab Chip 6(10):1274–1276
53. Lee GB, Lin CH, Chang GL (2003) Micro flow cytometers with buried SU-8/SOG optical waveguides. Sens Actuators A 103:165–170
54. Mariella R, vandenEngh G, Masquelier D, Eveleth G (1996) Flow-stream waveguide for collection of perpendicular light scatter in flow cytometry. Cytometry 24:27–31
55. Zhao HT, Yang Y, Chin LK, Chen HF, Zhu WM, Zhang JB, Yap PH, Liedberg B, Wang K, Wang G, Ser W, Liu AQ (2016) Optofluidic lens with low spherical and low field curvature aberrations. Lab Chip 16:1617–1624
56. Lim JM, Urbanski JP, Thorsen T, Yang SM (2011) Pneumatic control of a liquid-core/liquid-cladding waveguide as the basis for an optofluidic switch. Appl Phys Lett 98:044101
57. Hakanson U, Measor P, Yin D, Lunt E, Hawkins AR, Sandoghdar V, Schmidt H (2007) Tailoring the transmission of liquid-core waveguides for wavelength filtering on a chip. In: Proceeding of SPIE, silicon photonics II, vol 6477, p 647715
58. Jiang G, Baig S, Wang MR (2014) Polymer waveguide with tunable optofluidic couplers for card-to-backplane optical interconnects. In: Proceeding of SPIE, optical interconnects XIV, vol 8991, p 89910R
59. Li XC, Wu J, Liu AQ, Li ZG, Soew YC, Huang HJ, Xu K, Lin JT (2008) Liquid waveguide based evanescent wave sensor integrated onto a microfluidic chip. Appl Phys Lett 93:193901
60. Campopiano S, Bernini R, Zeni L, Sarro PM (2004) Microfluidic sensor based on integrated optical hollow waveguides. Opt Lett 29:1894–1896
61. White IM, Zhu H, Suter JD, Hanumegowda NM, Oveys H, Zourob M, Fan X (2007) Refractometric sensors for lab-on-a-chip based on optical ring resonators. IEEE Sens J 7:28–35
62. Mizaikoff B, Young C, Charlton C, Temelkuran B, Dellemann G, Giovannini M, Faist J (2006) Trace sensing with miniaturized mid-infrared sensors. In: 5th IEEE conference on sensors 2006, pp 331–333
63. Zhao HT, Zhang Y, Liu PY, Yap PH, Ser W, Liu AQ (2019) Chemical reaction monitoring via the light focusing in optofluidic waveguides. Sens Actuators B Chem 280:16–23
64. Schmidt BS, Yang AHJ, Erickson D, Lipson M (2007) Optofluidic trapping and transport on solid core waveguides within a microfluidic device. Opt Express 15:14322–14334
65. Yang AHJ, Moore SD, Schmidt BS, Klug M, Lipson M, Erickson D (2009) Optical manipulation of nanoparticles and biomolecules in sub-wavelength slot waveguides. Nature 457:71–75
66. Soltani M, Lin J, Forties RA, Inman JT, Saraf SN, Fulbright RM, Lipson M, Wang MD (2014) Nanophotonic trapping for precise manipulation of biomolecular arrays. Nat Nanotechnol 9:448–452
67. Chronis N, Liu GL, Jeong K-H, Lee LP (2003) Tunable liquid-filled microlens array integrated with microfluidic network. Opt Express 11:2370–2378
68. Mishra K, Murade C, Carreel B, Roghair I, Oh JM, Manukyan G, van den Ende D, Mugele F (2014) Optofluidic lens with tunable focal length and asphericity. Sci Rep 4:6378
69. Song CL, Luong TD, Kong TF, Nguyen NT, Asundi AK (2011) Disposable flow cytometer with high efficiency in particle counting and sizing using an optofluidic lens. Opt Lett 36:657–659
70. Chin LK, Liu AQ, Lim CS, Lin CL, Ayi TC, Yap PH (2010) An optofluidic volume refractometer using Fabry-Pérot resonator with tunable liquid microlenses. Biomicrofluidics 4:024107

71. Lim J, Gruner P, Konrad M, Baret JC (2013) Micro-optical lens array for fluorescence detection in droplet-based microfluidics. Lab Chip 13:1472–1475
72. Liu HL, Shi Y, Liang L, Li L, Guo SS, Yin L, Yang Y (2017) A liquid thermal gradient refractive index lens and using it to trap single living cell in flowing environments. Lab Chip 17:1280–1286
73. Garcia A, Gomez M, Acosta E (2012) Optical pressure sensor based on the combined system of a variable liquid lens and a point diffraction interferometer. Appl Opt 51:905–911
74. Kenji U, Akihide H, Hiroko K, Tsuguo S, Takehiko K (2000) Thermal lens microscope. Jpn J Appl Phys 39:5316
75. Cheng K, Zhong XQ, Xiang AP (2012) Optical trapping of metallic Rayleigh particles by using coherently and incoherently combined beams. Acta Physica Sinica 61(7):74202
76. Prikulis J, Svedberg F, Kall M, Enger J, Ramser K, Goksor M, Hanstorp D (2004) Optical spectroscopy of single trapped metal nanoparticles in solution. Nano Lett 4(1):115–118
77. Ramser K, Enger J, Goksor M, Hanstorp D, Logg K, Kall M (2005) A microfluidic system enabling Raman measurements of the oxygenation cycle in single optically trapped red blood cells. Lab Chip 5(4):431–436
78. Buican TN, Smyth MJ, Crissman HA, Salzman GC, Stewart CC, Martin JC (1987) Automated single-cell manipulation and sorting by light trapping. Appl Opt 26(24):5311–5316
79. Jess PRT, Garces-Chavez V, Smith D, Mazilu M, Paterson L, Riches A, Herrington CS, Sibbett W, Dholakia K (2006) Dual beam fibre trap for Raman microspectroscopy of single cells. Opt Express 14(12):5779–5791
80. Paterson L, Papagiakoumou E, Milne G, Garces-Chavez V, Tatarkova SA, Sibbett W, Gunn-Moore FJ, Bryant PE, Riches AC, Dholakia K (2005) Light-induced cell separation in a tailored optical landscape. Appl Phys Lett 87(12):123901
81. Sun YY, Ong LS, Yuan XC (2006) Composite-microlens-array-enabled microfluidic sorting. Appl Phys Lett 89(14):141108
82. Hart SJ, Terray A, Leski TA, Arnold J, Stroud R (2006) Discovery of a significant optical chromatographic difference between spores of Bacillus anthracis and its close relative, Bacillus thuringiensis. Anal Chem 78(9):3221–3225
83. Shi YZ, Xiong S, Chin LK, Zhang J, Ser W, Wu J, Chen T, Yang Z, Hao Y, Liedberg B, Yap PH, Tsai DP, Qiu C-W, Liu AQ (2018) Nanometer-precision linear sorting with synchronized optofluidic dual barriers. Sci Adv 4:eaao0773
84. Shi YZ, Xiong S, Chin LK, Yang Y, Zhang JB, Ser W, Wu JH, Chen TN, Yang ZC, Hao YL, Liedberg B, Yap PH, Zhang Y, Liu AQ (2017) High-resolution and multi-range particle separation by microscopic vibration in an optofluidic chip. Lab Chip 17:2443–2450
85. Shi YZ, Xiong S, Zhang Y, Chin LK, Chen Y-Y, Zhang JB, Zhang TH, Ser W, Larson A, Hoi LS, Wu JH, Chen TN, Yang ZC, Hao YL, Liedberg B, Yap PH, Tsai DP, Qiu C-W, Liu AQ (2018) Sculpting nanoparticle dynamics for single-bacteria-level screening and direct binding-efficiency measurement. Nat Commun 9:815
86. Kawata S, Tani T (1996) Optically driven Mie particles in an evanescent field along a channeled waveguide. Opt Lett 21(21):1768–1770
87. Hole JP, Wilkinson JS, Grujic K, Helleso OG (2005) Velocity distribution of Gold nanoparticles trapped on an optical waveguide. Opt Express 13(10):3896–3901
88. Shi Y, Tao H, Chin LK, Zhang Y, Yap PH, Ser W, Qiu C-W, Liu AQ (2020) Optical potential-well array for high-selectivity, massive trapping and sorting at nanoscale. Nano Lett 20(7):5193–5200
89. Shi Y, Tao H, Nguyen KT, Zhang Y, Chin LK, Zhu T, Yu Y, Cai H, Yap PH, Liu PY, Xiong S, Zhang J, Qiu C-W, Chan CT, Liu AQ (2019) Nanophotonic array-induced dynamic behavior for label-free shape-selective bacteria sieving. ACS Nano 13:12070–12080
90. Leach J, Mushfique H, di Leonardo R, Padgett M, Cooper J (2006) An optically driven pump for microfluidics. Lab Chip 6(6):735–739
91. Terray A, Oakey J, Marr DWM (2002) Microfluidic control using colloidal devices. Science 296(5574):1841–1844

92. Neale SL, Macdonald MP, Dholakia K, Krauss TF (2005) All-optical control of microfluidic components using form birefringence. Nat Mater 4(7):530–533
93. Kelemen L, Valkai S, Ormos P (2006) Integrated optical motor. Appl Opt 45(12):2777–2780
94. Boyer D, Tamarat P, Maali A, Lounis B, Orrit M (2002) Photothermal imaging of nanometer-sized metal particles among scatterers. Science 297(5584):1160–1163
95. Sershen SR, Westcott SL, Halas NJ, West JL (2002) Independent optically addressable nanoparticle-polymer optomechanical composites. Appl Phys Lett 80(24):4609–4611
96. Schroll RD, Wunenburger R, Casner A, Zhang WW, Delville JP (2007) Liquid transport due to light scattering. Phys Rev Lett 98(13):133601
97. Hellman AN, Rau KR, Yoon HH, Bae S, Palmer JF, Phillips KS, Allbritton NL, Venugopalan V (2007) Laser-induced mixing in microfluidic channels. Anal Chem 79(12):4484–4492
98. Dijkink R, Ohl CD (2008) Laser-induced cavitation based micropump. Lab Chip 8(10):1676–1681
99. Gonzalez-Avila SR, Huang XH, Quinto-Su PA, Wu T, Ohl CD (2011) Motion of micrometer sized spherical particles exposed to a transient radial flow: attraction, repulsion, and rotation. Phys Rev Lett 107(7):074503
100. Huang XH, Quinto-Su PA, Gonzalez-Avila SR, Wu T, Ohl CD (2010) Controlled manipulation and in situ mechanical measurement of single Co nanowire with a laser-induced cavitation bubble. Nano Lett 10(10):3846–3851
101. Wu J, Nyborg WL (2008) Ultrasound, cavitation bubbles and their interaction with cells. Adv Drug Deliv Rev 60(10):1103–1116
102. Coleman AJ, Whitlock M, Leighton T, Saunders JE (1993) The spatial-distribution of cavitation induced acoustic-emission, sonoluminescence and cell-lysis in the field of a shock-wave lithotripter. Phys Med Biol 38(11):1545–1560
103. El-Ali J, Sorger PK, Jensen KF (2006) Cells on chips. Nature 442:403–411
104. Ramada Q, Chen Y, Miao JH, Cheong HW, Oh S, Choo A, Tan HL (2007) Surface functionality enhancement for on-chip human embryonic stem cell culture. Sens Actuators B 126:351–353
105. Hung PJ, Lee PJ, Sabounchi P, Lin R, Lee LP (2005) Continuous perfusion microfluidic cell culture array for high-throughput cell-based assays. Biotechnol Bioeng 89:1–8
106. Chen CS, Mrksich M, Huang S, Whitesides GM, Ingber DE (1997) Geometric control of cell life and death. Science 276:1425–1428
107. Takayama S, Ostuni E, LeDuc P, Naruse K, Ingber DE, Whitesides GM (2001) Laminar flows: subcellular positioning of small molecules. Nature 411:1016
108. Wheeler AR, Throndset WR, Whelan RJ, Leach AM, Zare RN, Liao YH, Farrel K, Manger ID, Daridon A (2003) Microfluidic device for single-cell analysis. Anal Chem 75:3581–3586
109. Chin LK, Yu JQ, Fu Y, Yu T, Liu AQ, Luo KQ (2011) Production of reactive oxygen species in endothelial cells under different pulsatile shear stresses and glucose concentrations. Lab Chip 11:1856–1863
110. Nevill JT, Cooper R, Dueck M, Breslauer DN, Lee LP (2007) Integrated microfluidic cell culture and lysis on a chip. Lab Chip 7:1689–1695
111. Gascoyne PRC, Vykoukal JV (2004) Dielectrophoresis-based sample handling in general-purpose programmable diagnostic instruments. Proc IEEE 92:22–42
112. Fu AY, Chou H, Spence C, Arnold FH, Quake SR (2002) An integrated microfabricated cell sorter. Anal Chem 74:2451–2457
113. Chang WC, Lee LP, Liepmann D (2005) Biomimetic technique for adhesion-based collection and separation of cells in a microfluidic channel. Lab Chip 5:64–73
114. Xiao K, Grier DG (2010) Multidimensional optical fractionation of colloidal particles with holographic verification. Phys Rev Lett 104:028302
115. Crapisi A, Lante A, Pasini G, Spettoli P (1994) Enhanced microbial cell lysis by the use of lysozyme immobilized on different carriers. Process Biochem 28:17–21
116. Gao J, Yin X, Fang Z (2004) Integration of single cell injection, cell lysis, separation and detection of intracellular constituents on a microfluidic chip. Lab Chip 4:47–52
117. El-Ali J, Gaudet S, Günther A, Sorger PK, Jensen KF (2005) Cell stimulus and lysis in a microfluidic device with segmented gas–liquid flow. Anal Chem 77:3629–3636

118. Li PCH, Harrison DJ (1997) Transport, manipulation, and reaction of biological cells on-chip using electrokinetic effects. Anal Chem 69:1564–1568
119. Lee SW, Tai YC (1999) A micro cell lysis device. Sens Actuators A 73:74–79
120. Li ZG, Liu AQ, Klaseboer E, Zhang JB, Ohl CD (2013) Single cell membrane poration by bubble-induced microjets in a microfluidic chip dagger. Lab Chip 13(6):1144–1150

Chapter 2
Optofluidic Microlens

2.1 Introduction

Microlenses play a crucial role in optical systems, often employed to refract and alter the path of light. For instance, they can collimate diverging light from an optical fiber, directing it into a resonator for efficient light coupling, or focus light to a point for localized fluorescent excitation. Various fluid actuation techniques have been explored to create optofluidic microlenses, including hydraulic pressure [1], redox surfactants [2], electrowetting [3] and stimuli-responsive hydrogels [4] etc. However, these methods typically offer a limited range of curvature adjustment, involve complex fabrication processes, and focus light perpendicular to the device plane. Operating in an out-of-plane configuration, these fluidic lenses hold promise as replacements for conventional fixed lenses in devices such as cell phones and digital cameras. In contrast, on-chip optofluidic microlenses, classified as in-plane lenses, have been developed [5–8]. These lenses can be integrated with microfluidic networks that perform sample analysis, eliminating the need for complex alignment between different optical layers. The optofluidic microlens can be adjusted by altering its geometry or refractive index. The lens geometry can be modified using various microfluidic techniques. When optofluidic microlenses use the curvature formed by the interface between two immiscible liquids or between liquid and air, the radius of curvature can be adjusted by balancing the applied pressure with the interfacial tension. The refractive index of the liquid in an optofluidic device can be changed to modify the phase, amplitude, and polarization of the light passing through it. Although the refractive index can be altered by different methods such as pressure, temperature, electrowetting, and magnetic or electric fields, the simplest approach is to fill the lens with different liquids. The concentration of a solution can also be used to control the refractive index. In optofluidic microlenses that use a refractive index gradient to bend light beams, such as a liquid gradient refractive index (L-GRIN) lens [9, 10], adjusting the diffusion between flow streams with different

S. Xiong and A.-Q. Liu, *Optofluidics: Fundamentals and Applications*,
Microfluidics and Nanophotonics: Science and Engineering 4,
https://doi.org/10.1007/978-981-95-6296-1_2

species concentrations can modify the concentration profile and, consequently, the refractive index profile.

The most straightforward and widely-used method for creating tunable optofluidic microlenses is to use co-flow focusing [5, 6]. Figure 2.1 illustrates a typical on-chip optofluidic microlens configuration. By adjusting the liquid flow rates, researchers can precisely control the curvature of these microlenses. The design features an optical fiber that is aligned in the same plane as the microlens. Optofluidic microlenses have several unique characteristics that set them apart from other microlenses. Firstly, they can be dynamically reconfigured with different lens profiles by adjusting the flow rates in real time. Secondly, various combinations of liquids can be used to achieve different refractive index contrasts between the core and side flow streams, resulting in different focal distances. Thirdly, the liquid interfaces between the laminar flows are optically smoothed when the microlenses are formed in a low Reynolds number regime. Lastly, the chip fabrication process is simplified to a single step, significantly reducing complexity in both fabrication and integration. Consequently, optofluidic microlenses hold great potential for on-chip optical detection and manipulation.

This chapter focuses on the development of optofluidic microlenses, beginning with a theoretical analysis of their focusing power determined by liquid selection and resultant lens profiles. The microfluidic design includes structural optimization of the expansion chamber and flow-rate-based tunability mechanisms. By applying these principles, the optofluidic microlens delivers tunable performance, enabling compatibility with a broad range of optical system specifications.

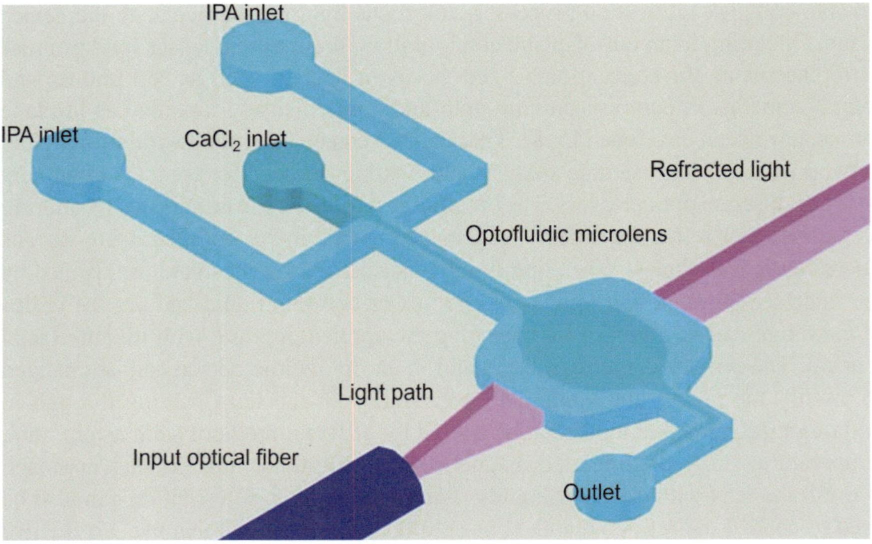

Fig. 2.1 Schematic illustration of the on-chip optofluidic microlens

2.2 Optofluidic Microlens via Laminar Flow

A typical optofluidic microlens is formed based on the hydrodynamic focusing flow with an expansion chamber. The flow field is governed by the incompressible Navier–Stokes equations, along with the continuity and momentum equations, which can be expressed as

$$\nabla_i u_i = 0 \tag{2.1}$$

and

$$\rho \frac{\partial u_i}{\partial t} + \rho\left(u_j \times \nabla_j\right)u_i = -\nabla_i p + \mu \nabla^2 u_i, \tag{2.2}$$

where u_i denotes the liquid velocity, μ is the dynamic viscosity, ρ and p are the density and the pressure, respectively. The symbols i and j represent spatial coordinate indices. When three flow streams are introduced into the focusing junction, a straight liquid interface forms between adjacent flows, with its position determined by the flow rate ratio. However, as these streams enter the expansion chamber, the interfaces between neighboring flows gradually shift from their original positions, as illustrated in Fig. 2.2. Similarly, upon exiting the expansion chamber, the interfaces gradually return to their initial positions. This creates curved interfaces in the expansion chamber, which function as a microlens when the liquids have differing refractive indices. The curvature direction and magnitude of the interface are position-dependent. A centrally located interface maintains a straight profile, while off-center positioning induces directional deflection. If positioned to the left, it deviates left in the expansion chamber, and vice versa.

By adjusting the flow rates of the streams, microlenses with various profiles and radii of curvature can be created, as depicted in Fig. 2.3. A biconvex microlens is formed by having higher flow rate in the central flow stream while keeping the side streams' flow rates constant ($Q_{\text{center}} > Q_{\text{left}} = Q_{\text{right}}$) as shown in Fig. 2.3a. Gradually increasing the flow rate of the right stream reduces the curvature, and when the right interface aligns with the center of the microchannel, it remains straight in the expansion chamber, resulting in a planar convex microlens, as illustrated in Fig. 2.3b. Further increasing the flow rate of the right stream shifts the right interface to the left of the microchannel's center, causing it to deviate left in the expansion chamber and forming a concave convex microlens, as shown in Fig. 2.3c.

The positions of the interfaces can be theoretically determined using the hydrodynamic focusing model as illustrated in Fig. 2.4. Assuming liquid A is injected into the central flow stream and liquid B into the side flow streams, the viscosity ratios and flow rate ratios between the side and central flow streams are defined as

$$\beta = \frac{\eta_B}{\eta_A} \tag{2.3}$$

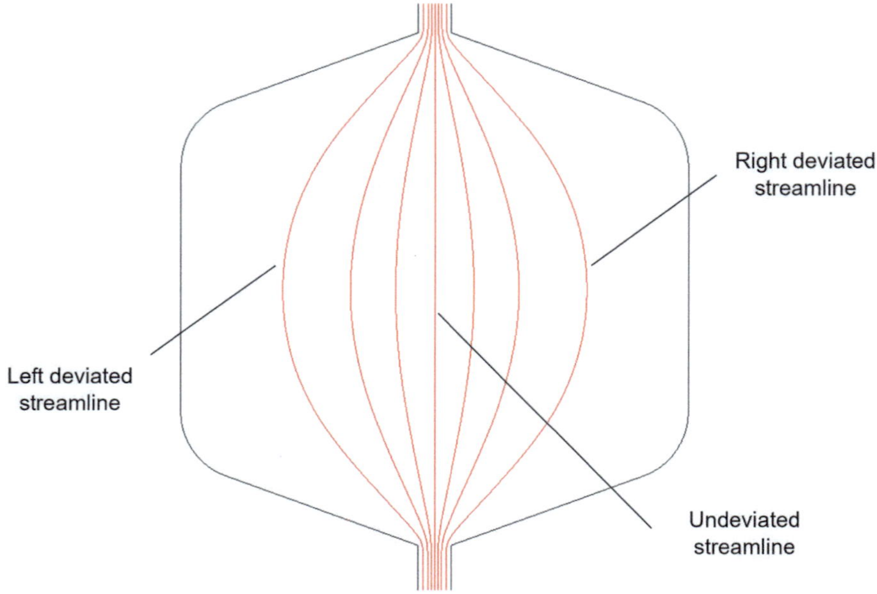

Fig. 2.2 Schematic illustration of the streamlines in an expansion chamber

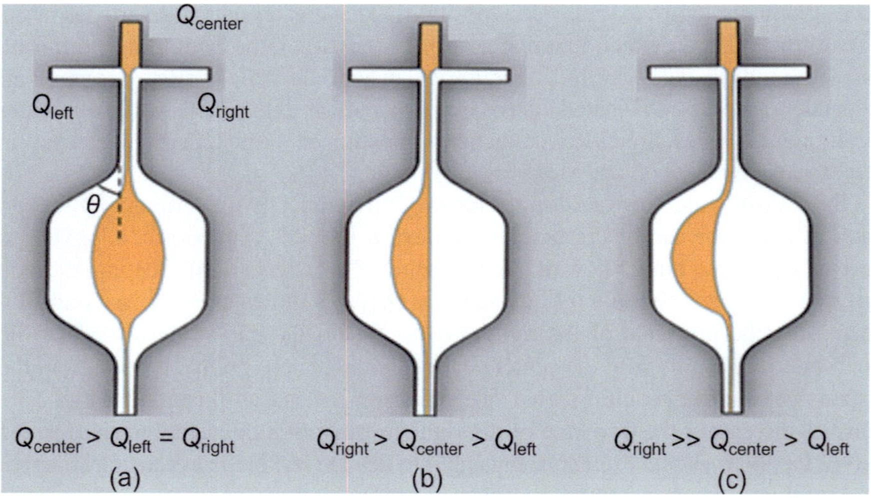

Fig. 2.3 Formation of optofluidic microlens: **a** biconvex microlens, **b** planar convex microlens, and **c** concave convex microlens. Reproduced from [6] with the permission of AIP Publishing

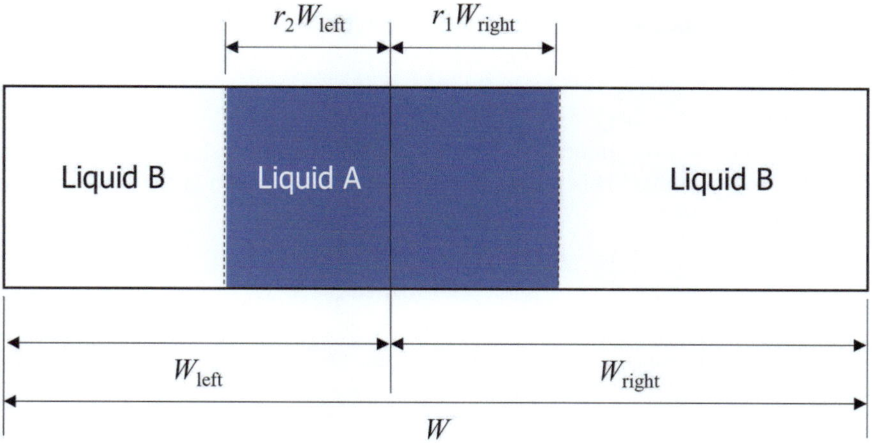

Fig. 2.4 Analytical model of flow stream interfaces in hydrodynamic focusing

and

$$\kappa_1 = \frac{Q_{right}}{Q_{center}} \quad \kappa_2 = \frac{Q_{left}}{Q_{center}}, \tag{2.4}$$

where η_A and η_B are the viscosities of liquid A and liquid B, respectively. The relative position of the core flow stream's centerline can be estimated as

$$W_{right} = \frac{W}{1 + (1 + 2\beta\kappa_2)/(1 + 2\beta\kappa_1)}. \tag{2.5}$$

The interface positions of the core flow stream can be determined using the following equations:

$$r_1 W_{right} = \frac{W_{right}}{1 + 2\beta\kappa_1}, \tag{2.6a}$$

and

$$r_2 W_{left} = \frac{W_{left}}{1 + 2\beta\kappa_2}. \tag{2.6b}$$

These equations are approximations, assuming the microchannel has a low aspect ratio.

The flow rate conditions of the three streams forming the optofluidic microlens and their respective lens profiles are analyzed. According to Eqs. (2.5) and (2.6a, b), a symmetrical biconvex microlens can only be formed when the flow rate ratios κ_1 and κ_2 are equal. Under this condition, the centerline of the core flow stream is positioned at the center of the microchannel ($W/2$). Based on the simulated streamline of the microlens, the radius of curvature of the interface is determined. As the central flow rate increases, the radius of curvature decreases while the microlens thickness increases. The smallest achievable radius of curvature is approximately half the width of the expansion chamber.

For a planar convex microlens, one of the interfaces must be straight within the expansion chamber. This means the interface is positioned at the center of the microchannel so that

$$W_{\text{right}} - r_1 W_{\text{right}} = \frac{W}{2}. \tag{2.7}$$

By substituting Eqs. (2.5) and (2.6a) into Eq. (2.7), the flow conditions can be defined as

$$\kappa_1 = \frac{1}{\beta} + \kappa_2. \tag{2.8}$$

To create a planar convex microlens, the flow rate of the right stream must exceed that of the left stream, displacing the core flow stream to the left of the microchannel's centerline. Furthermore, a higher flow rate ratio yields a planar convex microlens with reduced thickness and increased radius of curvature. Based on Eq. (2.8), the flow conditions for forming interfaces with different curvatures can be easily defined as follows,

$$\begin{array}{l} \text{Convex} \\ \text{Planar} \\ \text{Concave} \end{array} \begin{cases} \kappa_1 < \frac{1}{\beta} + \kappa_2 \\ \kappa_1 = \frac{1}{\beta} + \kappa_2 \\ \kappa_1 > \frac{1}{\beta} + \kappa_2 \end{cases} \quad \text{for right interface} \tag{2.9a}$$

and

$$\begin{array}{l} \text{Convex} \\ \text{Planar} \\ \text{Concave} \end{array} \begin{cases} \kappa_2 < \frac{1}{\beta} + \kappa_1 \\ \kappa_2 = \frac{1}{\beta} + \kappa_1 \\ \kappa_2 > \frac{1}{\beta} + \kappa_1 \end{cases} \quad \text{for left interface.} \tag{2.9b}$$

The transformation of the lens profile from biconvex to concave-convex can be studied by keeping the flow rate ratio κ_2 constant while gradually increasing the flow rate ratio κ_1. As the flow rate of the right stream is progressively increased, the right interface's radius of curvature expands correspondingly. When the interface becomes flat, forming a planar convex lens, the radius of curvature is infinite. Higher flow rates in the right stream make the interface concave, producing a concave-convex lens with a shrinking radius of curvature. The left interface exhibits a decreasing radius of curvature as it undergoes compression from the right interface. Similar effects occur on the left interface when the flow rate of the left stream is increased while keeping the other two streams constant.

The above analysis assumes that the flow streams in the chamber are laminar flows with optically smooth interfaces. Fortunately, microfluidic flows typically fall within the laminar flow regime, naturally arising with low Reynolds numbers. Of all dimensionless numbers, the Reynolds number (Re) is frequently mentioned in connection with microfluidics, representing the relative importance of inertial forces to viscous forces. It is expressed as

$$\mathrm{Re} = \frac{\rho u L}{\mu}, \tag{2.10}$$

where ρ is the density, u is the average flow velocity, μ is the viscosity of the fluid, and L is the characteristic flow dimension. Virtually without exception, microfluidic devices operate at low Reynolds numbers, signifying that viscous forces dominate over inertial forces, thereby ensuring laminar and highly predictable flow patterns [11].

2.3 Design and Formation of Optofluidic Microlens

Geometrical Design

To achieve optimal optical precision in an optofluidic microlens, which depends on the lens curvature and liquid stability, the geometrical design of the expansion chamber is crucial. The shape of the optofluidic microlens can be analyzed by simulating the flows of three laminar streams, focusing on two parameters: the width-to-length aspect ratio and the injection angle of the chamber.

Significant optical aberration occurs when the curvature of the core-cladding interfaces deviates notably from the ideal lens curvature. To minimize optical aberration, simulations of the three laminar flow streams are conducted in expansion chambers with aspect ratios of 1:1, 2:1, and 1:2, as illustrated in Fig. 2.5a. In an expansion chamber with a 2:1 aspect ratio, the interface shape at the central region is perfectly parallel to the chamber wall (the radius of curvature, $R = \infty$), representing the worst deviation from the ideal lens profile. Conversely, with a 1:2 aspect ratio, the interface shapes closely resemble the ideal lens shape. However, this configuration creates two recirculation zones at the chamber walls' ends, leading to liquid flow instability and unwanted asymmetry in the core-cladding interfaces. Therefore, to minimize interface deviation from the ideal lens shape and avoid instability caused by recirculation zones, the expansion chamber should have a 1:1 aspect ratio.

The sudden widening of the microchannel in the expansion chamber leads to abrupt changes in the flow rates of the liquid streams, inducing turbulence and creating recirculation zones at all edges with 90° expansion angles, as shown in Fig. 2.5b. These recirculation zones can be avoided by designing the chamber with an injection angle smaller than 90°, such as 70° or 60°. Comparing the two cases, the interface shape with a 70° injection angle aligns better with the ideal lens. Therefore,

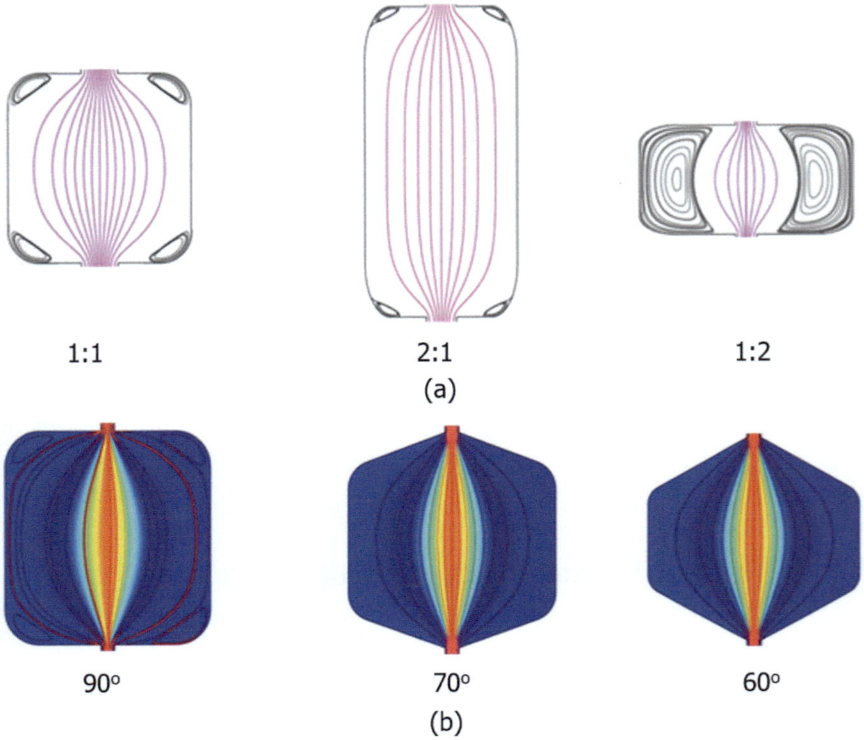

Fig. 2.5 **a** Simulated flows in the expansion chambers with width-to-length aspect ratio of 1:1, 2:1 and 1:2. **b** Simulated flows in the expansion chambers with injection angle of 90°, 70° and 60°

a 1:1 aspect ratio and a 70° injection angle are selected for the expansion chamber design.

Fabrication of PDMS Chip

The fabrication of the optofluidic chip utilizes the polymer soft lithography technique. The process involves patterning, developing, molding, and bonding, as illustrated in Fig. 2.6. Polydimethylsiloxane (PDMS) is an ideal material for microfluidic chips due to its biocompatibility, low cost, ease of fabrication, and suitability for mass production.

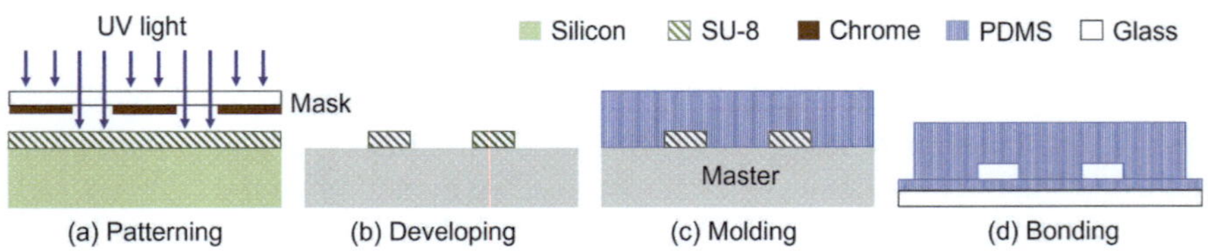

Fig. 2.6 Fabrication process of polymer soft lithography

The design layout of the optofluidic chip is created using computer-aided design (CAD) software (e.g., AutoCAD or L-Edit). A dark field chrome mask with clear features is commercially produced on a piece of sodalime glass based on the CAD drawing. To fabricate the master mold, a 6-inch silicon wafer is cleaned with acetone, isopropyl alcohol (IPA), and deionized (DI) water, then dried in an oven. A layer of SU-8 photoresist is spin-coated onto the wafer at 1800 rpm for 30 s. The photoresist-coated wafer is then soft baked at 65 °C for 6 min and 95 °C for 20 min to evaporate the solvent and densify the SU-8 layer. After exposure to UV light for 20 s under the sodalime glass mask, the wafer is baked at 65 °C for 1 min and 95 °C for 5 min. It is then developed by immersing it in SU-8 developer for 6 min. The substrate is rinsed with IPA and DI water, then dried with nitrogen. Finally, the wafer is hard baked at 120 °C in an oven for 10 min to further cross-link the SU-8 layer, completing the master fabrication.

Replica molding is an effective technique for duplicating the structure of a mold. The prepolymer of the PDMS elastomer is prepared by mixing the polymer base with the curing agent from the Sylgard 184 silicone elastomer kit (Dow Corning) at a 10:1 weight ratio. The pre-cured PDMS is degassed in a vacuum desiccator for 30 min. The mixture is then poured onto the master and degassed again for 15 min. The PDMS is cured at 75 °C for 1 h and then peeled off the master.

To complete a sealed microfluidic channel, the PDMS replica is bonded to another PDMS substrate. A flat layer of PDMS is coated on a glass slide and cured. A through hole is punched at each inlet and outlet of the PDMS replica using a puncher. Both PDMS pieces are oxidized by plasma treatment with a handheld corona treater for 3 min. The treated surfaces of the PDMS substrates bond spontaneously upon contact. For a stronger bond, the PDMS chip is annealed overnight, resulting in irreversible bonding.

Tuning of the Microlens Shape

The formation of biconvex microlenses with different radii of curvature is illustrated in Fig. 2.7a. In this example, two expansion chamber dimensions, 375 and 525 μm, are used to demonstrate the tunability of the microlens's radius of curvature. Liquids are injected into the optofluidic chip using syringe pumps. The side flow streams are maintained at a flow rate of 50 μL/min, with the flow rate ratio ($\kappa_1 = \kappa_2$) varying from 1 to 10. By gradually increasing the flow rate ratio, the radius of curvature increases while the microlens thickness decreases. The diffusive broadening effect is negligible with the selected liquids and flow rate, as shown in the insets. The smallest radii of curvature achieved by the 375 μm and 525 μm expansion chambers are 190 μm and 275 μm, respectively. Comparing these results, it can be concluded that smaller expansion chambers can lead to a larger tunable radius of curvature, as the minimum radius of curvature is limited by the expansion chamber's dimension (approximately half the width of the expansion chamber).

Figure 2.7b shows the experimental results of realizing three microlenses with varying lens profiles by adjusting the flow rate of the right flow stream. The expansion chamber used in the experiment measures 525 μm. By keeping the flow rate ratio between the left flow stream and the core flow stream at $\kappa_2 = 7$, the flow rate ratio

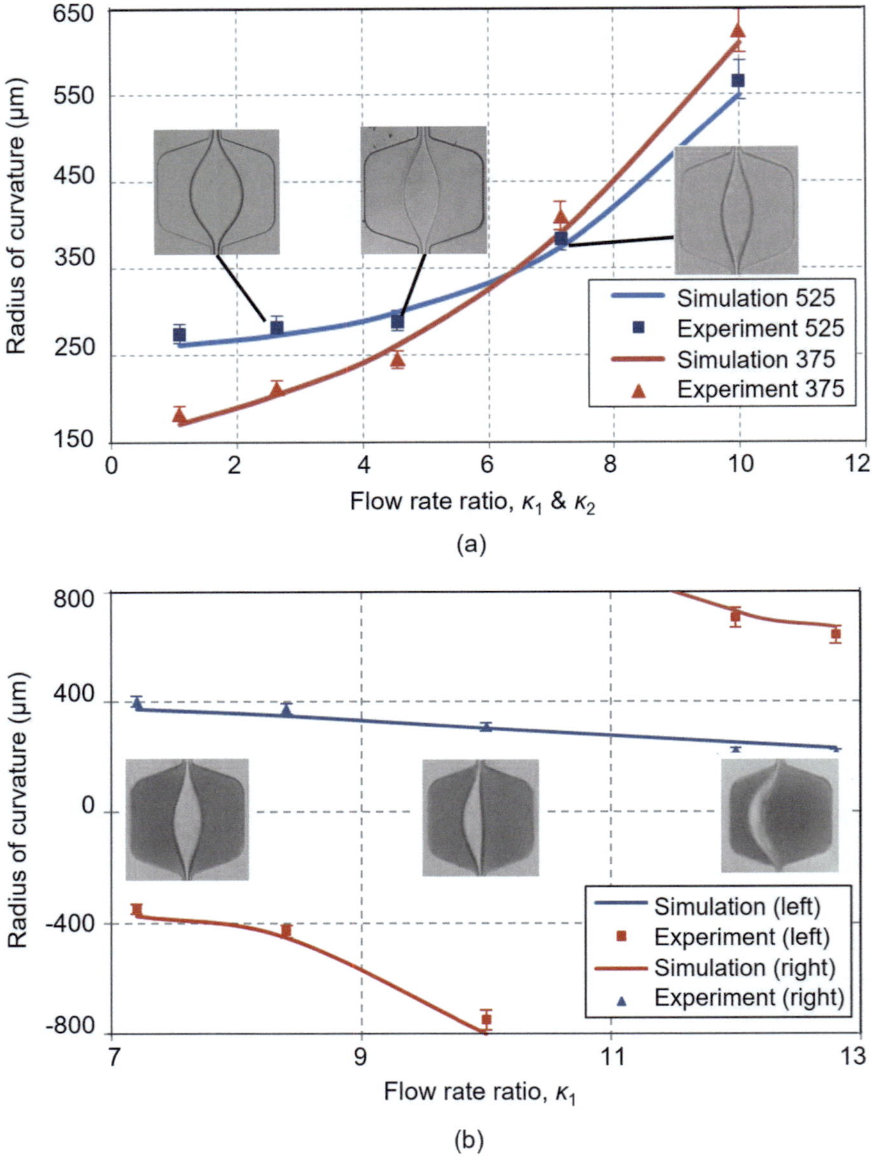

Fig. 2.7 **a** The radius of curvature in relation to changes in the flow rate ratio ($\kappa_1 = \kappa_2$). **b** The lens profile in relation to changes in the flow rate ratioκ_1

between the right flow stream and the core flow stream κ_1 is increased from 7 to 13. This adjustment allowed the transformation of the optofluidic microlens from a biconvex lens to a planar convex lens, and finally to a concave convex lens. Initially, a biconvex lens with a radius of curvature of 375 μm was formed in the chamber when both side flow streams had the same flow rate ratio. As the flow rate of the right flow stream increased, the radius of curvature of the right interface also increased, reaching infinity when the flow rate ratio κ_1 reached 11, resulting in the formation of a planar convex lens. Further increasing the flow rate of the right flow stream caused the right interface to bend towards the left, forming a concave convex lens. Throughout this process, the left interface was compressed by the right interface. The radius of curvature of the left interface decreased from 405 to 205 μm when the flow rate ratio κ_1 reached 13. Therefore, the use of laminar flows in the expansion chamber to form optofluidic microlenses offers flexibility in adjusting the radius of curvature of the interfaces and the shape of the microlens. The next section will explore and discuss the focusing power of the optofluidic microlens.

2.4 Focusing Power of Optofluidic Microlens

In lab-on-a-chip applications, microlenses are often used to collimate or focus diverging light, such as that emitted from an optical fiber. The focusing power of an optofluidic microlens can be calculated using the Lensmaker's formular,

$$\frac{1}{f} = (n-1)\left[\frac{1}{R_1} - \frac{1}{R_2} + \frac{(n-1)t}{nR_1R_2}\right], \tag{2.11}$$

where f represents the focal distance, n is the relative refractive index, t is the lens thickness, and R_1 and R_2 are the radii of curvature of the first and second interfaces, respectively. Figure 2.8 illustrates the focusing power of the microlenses. For a biconvex microlens, the focal distance decreases as the radius of curvature decreases as shown in Fig. 2.8a. To minimize propagation losses, a short focal distance is crucial, ensuring the distance between the light source or detector and the optofluidic microlens is minimized. Based on the Lensmaker's equation, the focal distance (f) for symmetrical biconvex microlenses ($R_1 = -R_2 = R$) can be determined as follows,

$$\frac{1}{f} = (n-1)\left[\frac{2}{R} + \frac{(n-1)t}{nR^2}\right]. \tag{2.12}$$

A shorter focal distance can be achieved by either increasing the relative refractive index or reducing the radius of curvature. To increase the relative refractive index, various combinations of liquids can be used for the central and side flow streams. In the simulation, a $CaCl_2$ solution with a refractive index of 1.462 and IPA with a refractive index of 1.382 are used, resulting in a refractive index contrast of 0.08.

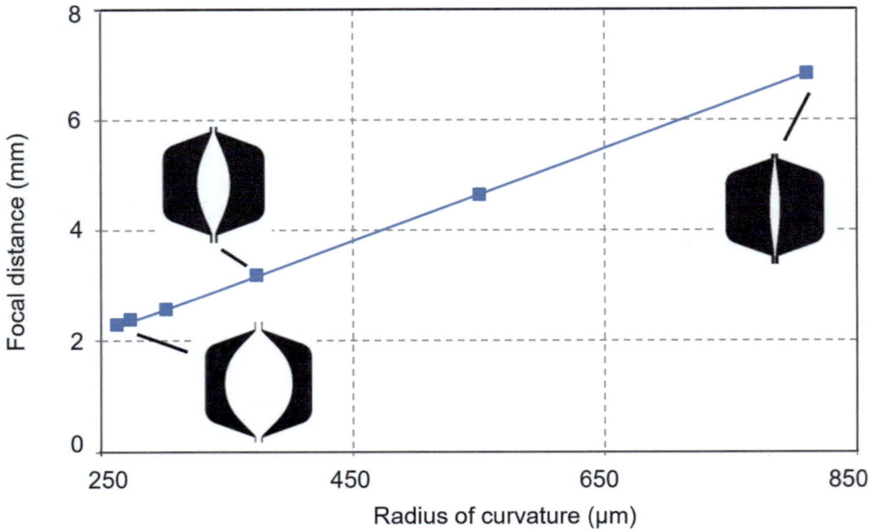

(a)

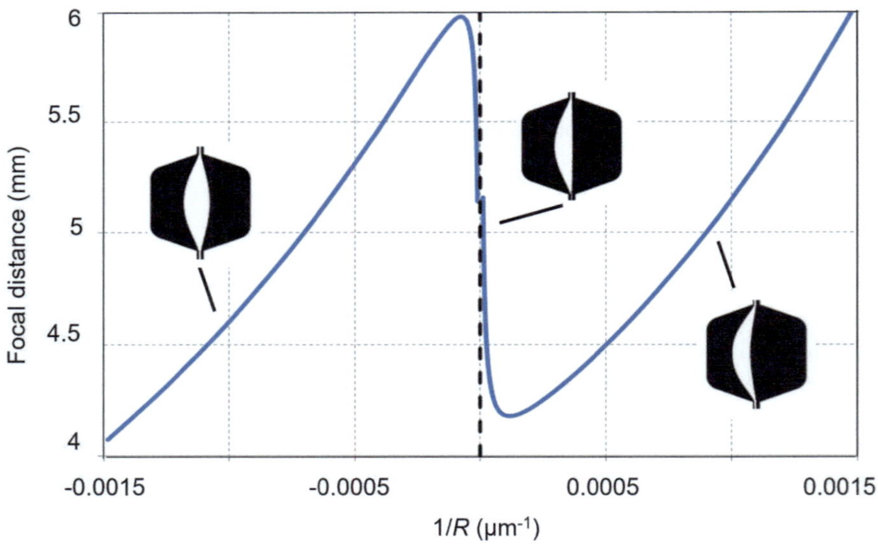

(b)

Fig. 2.8 Simulated focal distance of the optofluidic microlens with different **a** radius of curvature and **b** lens profiles

Additionally, a shorter focal distance can be obtained by designing a smaller chamber, as the minimum radius of curvature is constrained by the chamber's width (i.e., R_{min} = 1/2 width). However, a smaller chamber poses fabrication challenges because the microchannel is at least 15 times smaller than the chamber, and the beam width is limited by the chamber's length (aspect ratio 1:1). Therefore, the optical system's requirements must be considered during the design process.

Figure 2.8b illustrates the focusing powers of the microlens with various lens profiles. The focal distance increases when the right interface's radius of curvature changes from biconvex to planar convex. Conversely, the focal distance decreases near the planar convex microlens region because the left interface's radius of curvature is reduced as the right flow stream is compressed to the left. The left interface enhances the focusing power. However, when a concave interface forms on the right side without a significant reduction in the left interface's radius of curvature, the focal distance increases.

The tuning of the lens profile and the radius of curvature of the optofluidic microlens has been demonstrated. To form the liquid microlense, a 4.16 mol/L $CaCl_2$ solution with a refractive index of 1.462 and IPA with a refractive index of 1.382 are used as the core and cladding flow streams, respectively. A continuous-wave argon ion laser with a wavelength of 514 nm and an intensity of 150 mW is coupled into the optofluidic chip via a single-mode optical fiber. The focal distances for different microlenses are investigated. Figure 2.9a shows that the light emerged from the optical fiber is diverging without the formation of the optofluidic microlens. As the biconvex microlens forms in the expansion chamber, light propagating through it undergoes focusing, as illustrated in Fig. 2.9b. The focal distance is approximately 3.3 mm from the centerline of the expansion chamber. By analyzing the pixel intensity across the cross section, which is highlighted by the dotted lines, the focused light has a higher intensity of approximately threefold.

The tunability of the focal distance via the change of the flow rate ratio is investigated with the formation of optofluidic biconvex microlens in the expansion chamber. Figure 2.9c demonstrates the tunability of the focal distance with respect to the radius of curvature of the biconvex microlens. In the experiment, the expansion chamber has a dimension of 525 μm × 525 μm and the formed microlens has a minimum radius of curvature of 275 μm. The focusing power of the optofluidic microlens is limited to greater than 2 mm with the current liquid combination. By using a different liquid combination with a refractive index contrast of 0.27, the focusing power can be improved to approximately 800 μm.

Another effective method to shorten the focal distance is by using compound biconvex microlenses. The effective focal distance of these compound biconvex microlenses is estimated to be

$$\frac{1}{f} = \frac{1}{f_1} + \frac{1}{f_2} - \frac{d}{f_1 f_2}, \tag{2.13}$$

where d is the distance between the two microlenses, f_1 and f_2 represent the focal lengths of the first and second microlenses, respectively. For microlenses with a focal

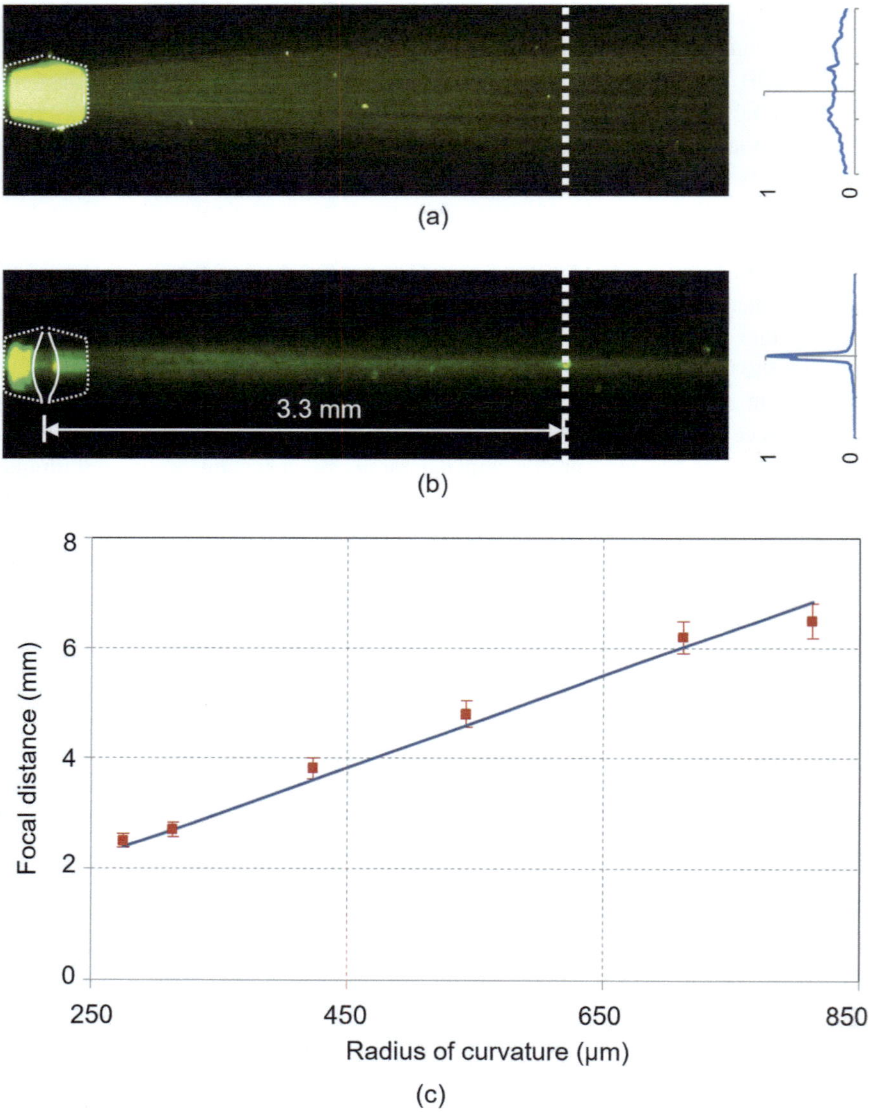

Fig. 2.9 Optical micrographs of **a** diverging light emitted from the optical fiber, and **b** focused light with the presence of the optofluidic microlens. **c** Relationship between biconvex microlens focal distance and the radius of curvature

length of 2 mm and a separation of 625 μm, the effective focal length is reduced to 1.185 mm.

In the compound microlenses experiment, the dimensions of the expansion chamber and the liquid selection are identical to those used for a single microlens. The two optofluidic microlenses are separated by a 100-μm PDMS gap between the expansion chambers, as illustrated in Fig. 2.10a. When light passes through these microlenses, it undergoes two refractions and is subsequently focused on a point closer than with a single microlens. As shown in Fig. 2.10b, the second microlens maintains a radius of curvature of 300 μm, while the radius of curvature of the first microlens can be adjusted from 275 to 370 μm by changing the flow rate ratio. The focal distance is defined as the distance from the focal point to the centerline of the first expansion chamber. The observed change in focal distance aligns with predictions made using Eq. (2.13). The focal distance has been reduced from 2.5 to 1.4 mm. To further decrease the focal distance, the second microlens can be adjusted to achieve a smaller radius of curvature.

In summary, there are three effective methods to enhance focusing power by reducing the focal distance. First, a smaller expansion chamber dimension results in a smaller achievable radius of curvature, thereby reducing the focal distance. Second, altering the liquid combination among the three flow streams can increase the refractive index contrast, leading to higher focusing power. Third, using compound microlenses can shorten the focal distance. When two optofluidic microlenses are employed, the light ray experiences double refraction, resulting in a focal point shorter than that achieved with a single microlens.

2.5 Optofluidic Microlenses for Light Collimation

A biconvex microlens has greater focusing power or a shorter focal distance compared to other types of microlenses. Consequently, using a biconvex microlens for light focusing or collimation can reduce the spatial distance of the optical system and the size of the PDMS chip. Additionally, forming a biconvex microlens is easier and more robust because the flow rates of both cladding flow streams are equal, utilizing a single inlet and a single syringe pump. In the following example, a pair of optofluidic microlenses provide real-time tunable light collimation for an on-chip resonator. A gold-coated microchannel between the liquid microlenses confines and resonates the light.

Figure 2.11a illustrates the formation of liquid microlenses when the core flow stream is maintained at a flow rate of 35 μL/min. By using a faster flow rate, the diffusion at the liquid interfaces due to the concentration difference between the two liquid layers is minimized, as the diffusion distance of Ca^{2+} and Cl^{-} ions is reduced. With the selected liquids, the diffusive broadening of the interfaces is eliminated, resulting in microlenses that closely match the ideal lens. Additionally, the liquid microlenses have optically smooth curvatures, ensuring high-quality light manipulation and collimation.

The light propagation path is illustrated in Fig. 2.11b. The first liquid microlens collimates the diverging beam, which is then coupled back into the optical output fiber by the second liquid microlens. Figure 2.12 illustrates collimation efficiency relative to the microlenses' curvature radius. Without collimation, only a small portion of light (parallel light at the central beam) resonates within the FP cavity, while diverging

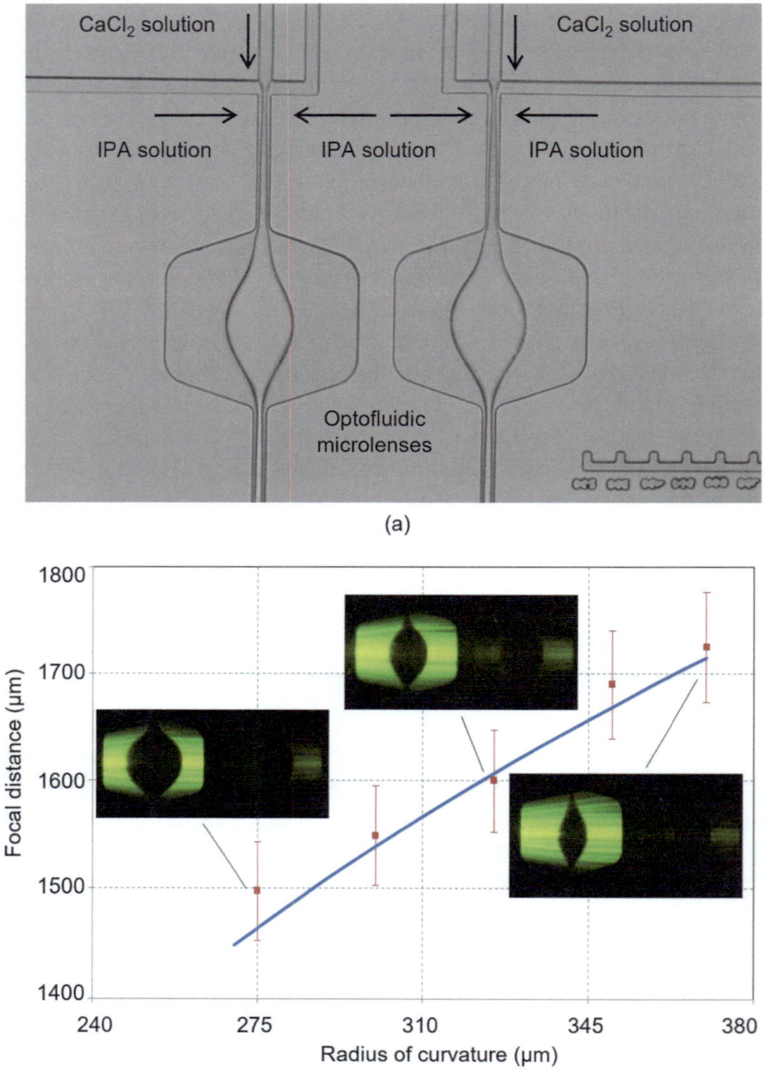

(a)

(b)

Fig. 2.10 **a** Micrograph of the compound optofluidic microlenses. **b** Change in focal distance of the combinational microlenses with the variation in the radius of curvature of the first microlens

light is observed outside the FP cavity in the microchannel, as shown in Fig. 2.12a. However, when a liquid microlens with a radius of 150 μm is formed, more light is collimated and resonates, as shown in Fig. 2.12b. Less diverging light is observed outside the FP cavity in the microchannel. When a liquid microlens with a radius of 138 μm is formed, the light emerging from the optical fiber is fully collimated, as shown in Fig. 2.12c. It is evident that most light is collimated and resonates within

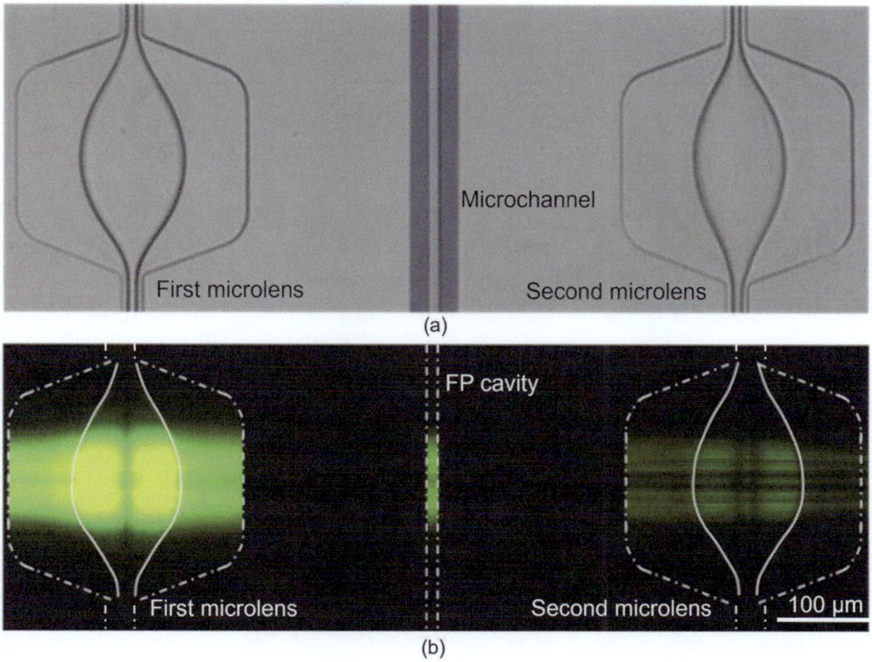

Fig. 2.11 Micrograph of **a** the optofluidic microlenses and **b** the light propagation path. Reproduced from [12] with the permission of AIP Publishing

the FP cavity, with nearly negligible diverging light observed outside the FP cavity in the microchannel. The light beam appears parallel when observed using a 40 × objective lens, as shown in Fig. 2.12d.

Optical resonators are engineered to attain higher finesse, thereby enhancing measurement resolution and precision. A well-collimated light beam contributes to improving the resonator's finesse. In this example, the resonant cavity's microchannel, which is 15 μm wide, is filled with DI water. The finesse of the FP cavity significantly improves from 3.5 to 18.8 when liquid microlenses with a curvature radius of 138 μm are formed. Under these conditions, most of the light is collimated, resonates, and efficiently couples back to the detection fiber. If the curvature radius of the liquid microlenses is further reduced, the light from the optical fiber converges, making it inefficient for resonance. Consequently, the finesse of the FP cavity decreases. To maintain a finesse higher than 15, the microlenses should have curvature radii between 130 and 145 μm.

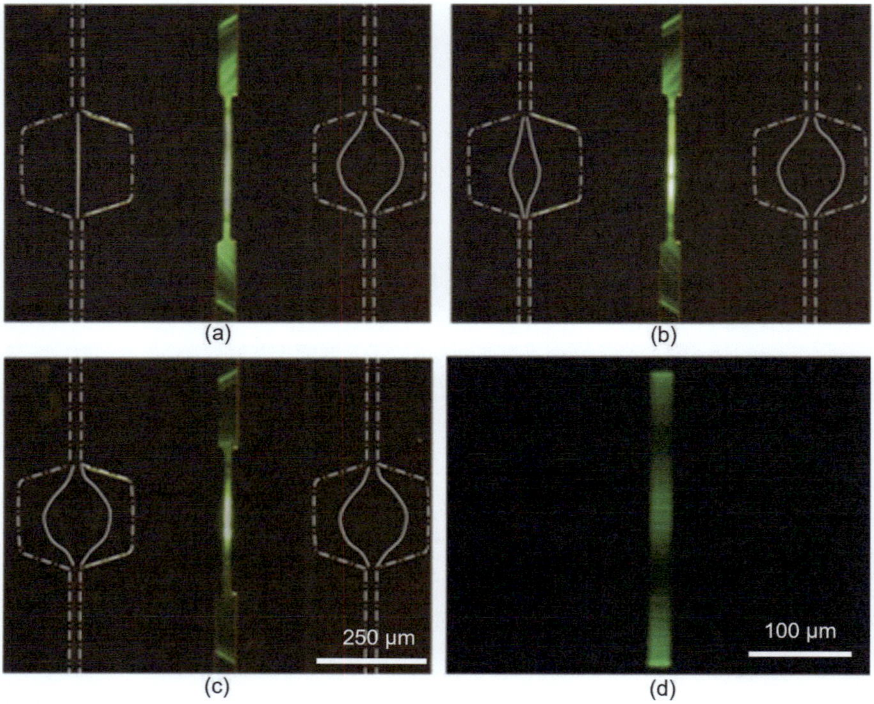

Fig. 2.12 Optical micrographs of the light beam when **a** no microlens, **b** microlens with a radius of 150 μm, and **c** microlens with a radius of 138 μm is formed. **d** Zoom view of the collimated beam in the cavity. Reproduced from [12] with the permission of AIP Publishing

2.6 Summary

Optofluidic microlenses are considered as lenses that use liquid as the optical medium and microfluidic techniques as a means for actuation and manipulation. Hydrodynamic focusing is a straightforward method for creating optofluidic microlenses. By carefully designing the lens chamber and positioning the inlets and outlets, it is possible to create a complex lens system with both concave and convex surfaces. This chapter demonstrates real-time tuning of the optofluidic microlens and investigates its focusing power. The combined microlenses achieve a shorter focal distance and offer a broader tuning range. The interface between two laminar flows, which forms the curvature of the optofluidic microlens, remains optically smooth. Faster flow rates minimize diffusion at the liquid interfaces caused by concentration differences between the liquid layers, reducing the diffusion distance of Ca^{2+} and Cl^{-} ions. The optical tunability depends on the microlens's tunability and the dynamic modification of optical properties through local variations in refractive indices. In the described microlens, a refractive index contrast of 0.08 is achieved, which is significantly higher than that of thermal and electro-optic effects [3]. Future studies may explore using other physical fields, such as temperature and electric fields, to modulate the refractive index without changing the liquids. Advanced flow control components like microvalves and micropumps could further enhance control precision.

References

1. Zhang DY, Lien V, Berdichevsky Y, Choi J, Lo YH (2003) Variable-focus liquid microlenses and microlens arrays actuated by thermoresponsive hydrogels. Appl Phys Lett 82:3171–3172
2. Ren H, Wu ST (2005) Tunable electronic lens using a gradient polymer network liquid crystal. Appl Phys Lett 86:211107
3. Kuiper S, Hendriks BHW (2004) Variable-focus liquid lens for miniature cameras. Appl Phys Lett 85:1128–1130
4. Dong L, Agarwal AK, Beebe DJ, Jiang H (2006) Adaptive liquid microlenses activated by stimuli-responsive hydrogels. Nature 442:551–554
5. Tang SKY, Stan CA, Whitesides GM (2008) Dynamically reconfigurable liquid-core liquid-cladding lens in a microfluidic channel. Lab Chip 8:395–401
6. Seow YC, Liu AQ, Chin LK, Li XC, Huang HJ, Cheng TH, Zhou XQ (2008) Different curvatures of tunable liquid microlens via the control of laminar flow rate. Appl Phys Lett 93:084101
7. Nguyen N-T (2010) Micro-optofluidic lenses: a review. Biomicrofluidics 4:031501
8. Chen Q, Li T, Li Z, Long J, Zhang X (2018) Optofluidic tunable lenses for in-plane light manipulation. Micromachines 9(3):97
9. Mao X, Lin SC, Lapsley MI, Shi J, Juluri BK, Huang TJ (2009) Tunable Liquid Gradient Refractive Index (L-GRIN) lens with two degrees of freedom. Lab Chip 9:2050–2058
10. Zhao HT, Yang Y, Chin LK, Chen HF, Zhu WM, Zhang JB, Yap PH, Liedberg B, Wang K, Wang G, Ser W, Liu AQ (2016) Optofluidic lens with low spherical and low field curvature aberrations. Lab Chip 16:1617–1624
11. Squires TM, Quake SR (2005) Microfluidics: fluid physics at the nanoliter scale. Rev Mod Phys 77(3):977–1026
12. Chin LK, Liu AQ, Lim CS, Lin CL, Ayi TC, Yap PH (2010) An optofluidic volume refractometer using Fabry-Pérot resonator with tunable liquid microlenses. Biomicrofluidics 4:024107

Chapter 3
Optofluidic Prism

3.1 Introduction

A prism is an essential element for adjusting optical paths, such as directing light beams and controlling dispersion, and is widely used in optical analysis systems. Liquid prisms have been developed since 1938. When a rectangular trough filled with liquid is rotated around one of its lower edges, the liquid forms a prism with an angle that varies with the rotation [1]. This type of liquid prism is used to construct an auto-collimating microscope, significantly improving the accuracy of refractive index measurements. Another type of liquid prism consists of liquid or gas enclosed in a sector-shaped bag with two transparent, stiffened side walls (120 $\times$ 80 mm) [2]. The apex angle of this prism can be adjusted by changing the distance between the bases of the two walls, and its refractive index can be modified by altering the medium inside the bag.

In recent years, the size of prisms has been decreasing due to the miniaturization of optical systems. A micro-liquid prism has been developed by placing two transparent plates on a liquid droplet and encapsulating the structure with parylene [3]. Surface tension automatically positions the two plates, which range from 400 to 1200 μm in diameter. Once the prism's shape is fixed by encapsulation, its refraction angle becomes unchangeable. This prism has been validated in surface plasmon resonance (SPR) measurements. Another small prism, consisting of two plates (2 to 4 mm in diameter) and a sandwiched liquid layer (1.5 mm high), has been developed to allow wedge angle tuning via electrowetting [4]. The wedge angle increases with applied voltage, reaching a maximum of 13° at 100 V. This prism can shift the light path by 8° and the field of view by 3°. However, it cannot be maintained for long periods due to volume changes from evaporation. These micro-prisms are all out-of-plane optical components.

On-chip micro-prisms are typically fabricated using PDMS, either with or without liquid filling in the hollow prism [5–7]. Early studies showed that the light beam

S. Xiong and A.-Q. Liu, *Optofluidics: Fundamentals and Applications*,
Microfluidics and Nanophotonics: Science and Engineering 4,
https://doi.org/10.1007/978-981-95-6296-1_3

shifted before and after oil was added to the prism chamber [5]. Subsequently, optofluidic prisms have been employed for on-chip detection. A hollow Abbe prism has been utilized for absorption detection and refractive index measurement of chemical species within it [6]. The wavelength shift collected by the output optical fiber can be used to calculate variations in the refractive index. The total volume of the prism is 1.579 μL, and the minimum detectable fluorescein concentration is 112.5 μM. By adjusting the volume and geometry, the performance of the prism has been optimized, achieving a detection limit of 1.83 μM for fluorescein using a prism with a total deviation angle of 90° [7]. These prisms can be designed on a microchip with a high degree of monolithic integration, although their rigid geometry poses challenges for the development of optofluidic systems. Later, a tunable symmetrical prism with a liquid-core liquid-cladding structure was developed in a sector-shaped chamber [8]. This prism is formed by three laminar flows, and its apex angle is adjusted by varying the flow rates. The chamber functions as a source-sink pair within a divergent boundary with a 90° opening angle. Consequently, the prism exhibits a tunable apex angle that is always less than 90°. When the chamber is completely filled with the core liquid, the deviation angle reaches a maximum of 11.4°. Expanding the tuning range of the apex angle could result in a larger deviation angle. Conversely, another optofluidic prism structure, based on two laminar flows, features a tunable apex angle that is always greater than the chamber angle [9]. This chapter presents a case study of an optofluidic prism, focusing on its design principles, geometric and optical tunability, and applications in light steering and chemical analysis.

3.2 Shape Control of Tunable Prism

A tunable optofluidic prism is created using two laminar flow streams with different refractive indices, as illustrated in Fig. 3.1. These streams are injected into a triangular chamber, with the inlet and outlet connected to the two base corners of the chamber. The inner flow stream forms the main part of the prism, while the outer flow stream adjusts its shape. By controlling the flow rates of these streams, the apex angle of the optofluidic prism can be tuned, thereby altering the deviation angle of the incident light beam. Additionally, a third channel at the vertex of the triangular chamber injects a control flow. The same solutions are used in both the outer and control flow streams. This control flow allows the prism's shape to be modified between symmetrical and asymmetrical configurations, as shown in Fig. 3.1b–d.

The incident light is directed into a single-mode optical fiber aligned parallel to the base of the prism. A convex lens positioned between the fiber and the prism collimates the light beams before they enter the prism. On the opposite side of the prism, a semicircular chamber filled with fluorescent dye solution is placed nearby. This dye can be excited by the incident light, allowing it to trace the output light beams.

The optofluidic prism's geometry can be theoretically modeled using hydrodynamic control principles. The interface position between co-flowing streams in

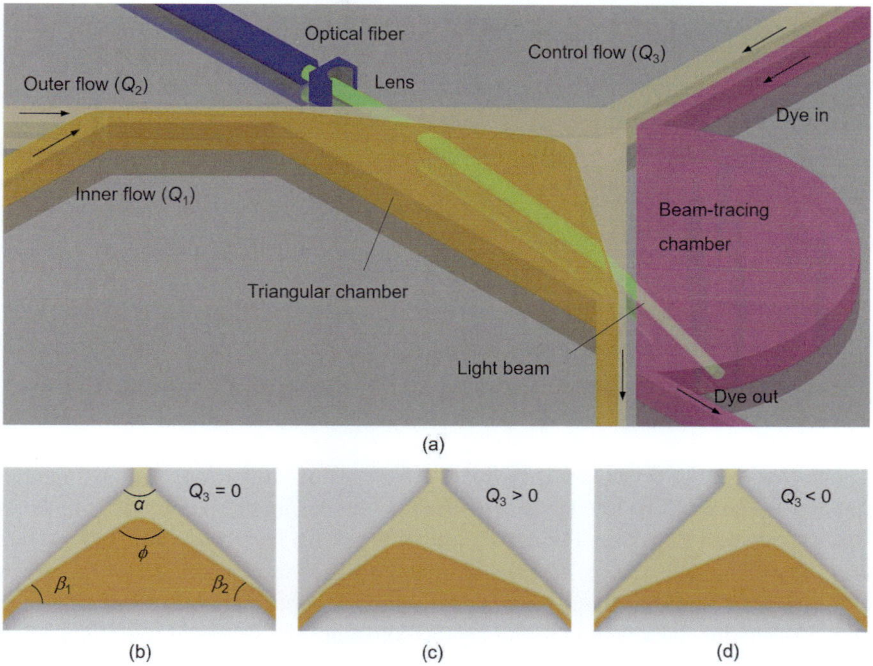

Fig. 3.1 **a** Schematics of the optofluidic prism. The shape is adjusted by varying the flow rates: **b** symmetric prism, **c** left-shifting asymmetric prism, and **d** right-shifting asymmetric prism. Reproduced from [9] with permission from the Royal Society of Chemistry

the microchannel follows convective-diffusive transport theory [10, 11]. For a flat channel ($h \ll w$), the interface position of the two streams can be determined based on mass conservation as follows

$$\gamma = \frac{w_1}{w} = \frac{1}{1 + \frac{\eta_2}{\eta_1} \cdot \frac{Q_2}{Q_1}}, \tag{3.1}$$

where w_1 is the width of the inner flow and w is the width of the channel. The viscosities of the inner and outer liquids are denoted by η_1 and η_2, respectively. Similarly, Q_1 and Q_2 are the flow rates of the inner and outer flows, respectively.

The interface between the two flow streams within the chamber is theoretically a straight line extending from the inlet to the vertex of the chamber. The distribution of the flow flux is proportional to the width ratio of the flow streams. Consequently, a function of the base angle β_1 that addresses the ratio between Q_1 and Q_2 can be derived as

$$\beta_1 = \frac{90^\circ - \frac{\alpha}{2}}{1 + \frac{\eta_2}{\eta_1} \cdot \frac{Q_2}{Q_1}}, \tag{3.2}$$

where α is the apex angle of the triangular chamber. In this design, a control flow, which is the same liquid as the outer flow with a flow rate Q_3, is introduced to shape the prism. Consequently, the other base angle β_2 is influenced not only by Q_1 and Q_2, but also by Q_3. The value of Q_3 is positive when the control flow is injected into the chamber and negative otherwise. Therefore, the base angle β_2 can be expressed as

$$\beta_2 = \frac{90^\circ - \frac{\alpha}{2}}{1 + \frac{\eta_2}{\eta_1} \cdot \frac{Q_2+Q_3}{Q_1}}. \tag{3.3}$$

The apex angle of the prism is written as

$$\phi = 180^\circ - \beta_1 - \beta_2. \tag{3.4}$$

The shape of the prism can vary under different flow conditions, as illustrated in Fig. 3.1. When the flow rate Q_3 is zero, the prism is symmetric about the chamber's axis, as shown in Fig. 3.1a. The apex angle of the prism increases with the flow rate ratio of Q_2 to Q_1. When Q_1 and Q_2 are kept constant, the apex angle of the prism increases with the flow rate Q_3, causing the prism to appear asymmetrical. The base angle β_1 remains fixed when Q_1 and Q_2 are constant. The vertex of the prism shifts due to the difference between the base angles β_1 and β_2. When Q_3 is positive, β_2 is smaller than β_1, resulting in the vertex of the prism being on the left side of the chamber, as shown in Fig. 3.1b. Conversely, when Q_3 is negative, β_2 is larger than β_1, and the vertex of the prism shifts to the right side of the chamber, as shown in Fig. 3.1c.

To achieve an optofluidic prism with optimal optical precision and tuning range, the chamber's structure must be optimized. Since the optofluidic prism is formed by two laminar flow streams, the stability of these flows is crucial for its performance. COMSOL Multiphysics is used to simulate fluid behavior with various geometrical designs of different chambers. The prism shape is analyzed by simulating the flow streams in chambers with different apex angles. According to the simulation results, a smaller apex angle allows for a larger variation in the prism's apex angle. For instance, within the same flow rate ratio range ($Q_2/Q_1 = 0$–5), the tuning range of the apex angle is 50° for a chamber with a 110° apex angle, whereas it can be 100° for a chamber with a 50° apex angle. However, recirculation zones caused by turbulent flow occur at the chamber's vertex when the apex angle is smaller than 90°, as shown in Fig. 3.2a. This results in a flatter top of the optofluidic prism, reducing its effective working length. To balance these requirements, chambers with apex angles of 70° and 90° are selected for the experiments.

The effect of inlet width was analyzed based on simulation results for a chamber with a 90° apex angle. The inlet width to base length ratios were set at 1:22 and 1:8. For optofluidic components, it is essential to maintain a smooth and clear interface between flows. Figure 3.2b demonstrates that narrower inlets promote diffusion between flows, indicating that overly narrow inlets are not ideal for forming an

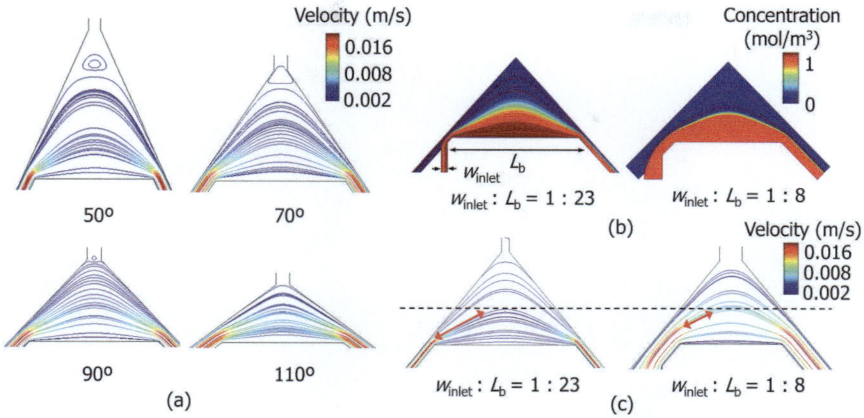

Fig. 3.2 Simulation results for **a** flow streams in various chamber shapes, **b** diffusion with different inlet width to base length ratios, and **c** flow streams with varying inlet width to base length ratios. The arrowed line indicates the effective working length for each chamber when the prism height is constant. Reproduced from [9] with permission from the Royal Society of Chemistry

optofluidic prism. Moreover, it is necessary to investigate the relationship between inlet width and working length, defined as the straight-line segment of the interface from the inlet to the prism's vertex. As shown in Fig. 3.2c, wider inlets result in shorter working lengths. After considering both diffusion and working length, the optimal ratio of inlet width to base length was determined to be 1:18.

The prism's shape is influenced by the laminar flow concept, which greatly depends on the channel size. As the chamber size increases, the diffusion between the two flows also increases. However, as the chamber size decreases, both the working distance and the length of the optical path are reduced. Consequently, the tuning range of the optofluidic prism is limited. To fully capture the experimental data, the field of view size in the microscope system must also be considered. Ultimately, the optimal design is determined to be $L_b = 3600$ μm and $w_{inlet} = 200$ μm.

The optofluidic chip is fabricated using the polymer soft lithography technique. The channel height is approximately 100 μm to allow for the insertion of an optical fiber, which serves as the light source. A PDMS lens is positioned between the fiber insertion channel and the optofluidic prism chamber to collimate the light beam. Additionally, a semicircular chamber is created on the right side of the prism chamber to trace the output light beam, as illustrated in Fig. 3.3.

Benzyl alcohol ($n = 1.540$, viscosity $\eta = 5.47$ mPa s at 25 °C) and DI water ($n = 1.332$, viscosity $\eta = 0.89$ mPa s at 25 °C) are used as the inner liquids. An outer liquid mixture, consisting of 80% ethylene glycol and 20% DI water with a viscosity of approximately 8 mPa s at 25 °C, has an effective refractive index matching that of PDMS ($n = 1.412$). This refractive index matching reduces light scattering on the microchannel wall. The mixture of ethylene glycol and DI water ($n = 1.412$), containing rhodamine 6G and fluorescent red 646, is injected into the beam tracing chamber to visualize the optical path. Rhodamine 6G and fluorescent red 646 are

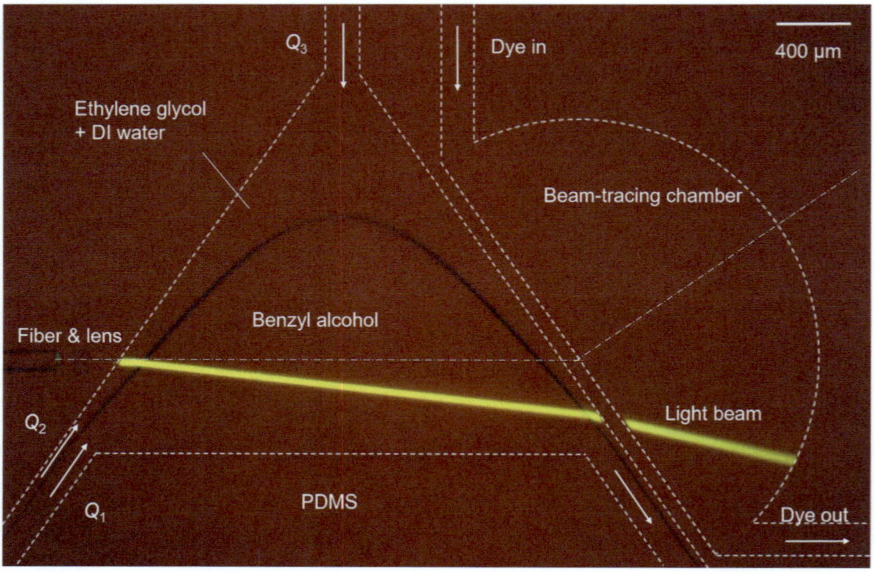

Fig. 3.3 Photograph of the optofluidic prism. Reproduced from [9] with permission from the Royal Society of Chemistry

excited by 488-nm and 633-nm lasers, respectively. To easily image the shape of the optofluidic prism, rhodamine 6G is added to the inner liquid, and a fluorescent image is captured.

To validate the theory predicting the relationship between flow rates and the apex angle of the optofluidic prism, a chamber with a 70° apex angle was utilized. The inner flow, containing benzyl alcohol with dissolved fluorescent dye, was excited and appeared red. The outer flow, filled with an ethylene glycol solution ($n = 1.412$), appeared black in the image. The interface between the dark and red regions was clearly visible, as shown in Fig. 3.4.

For a symmetric prism, the control flow rate is zero. The outer flow rate is fixed at 10 μl/min, while the inner flow rate varies from 10 to 180 μl/min to adjust the apex angle of the optofluidic prism, as shown in Fig. 3.4a. Consequently, the apex angle changes from 135° to 75°. As the inner flow rate increases, the interface between the two liquids moves closer to the chamber's sidewalls, resulting in a smaller apex angle. The relationship between the flow rates and the apex angle of the prism is analytically estimated using Eqs. (3.2)–(3.4) and is represented by the solid line in the Fig. 3.4a, which aligns well with the theoretical prediction. Without the influence of the control flow, a symmetric shape is maintained throughout the flow rate variation process. However, the prism shape is not perfect, with the top of the interface forming an arc shape, as noted in the simulation results. To minimize the impact of this deviation on light propagation, the light insertion is positioned closer to the lower part of the optofluidic prism.

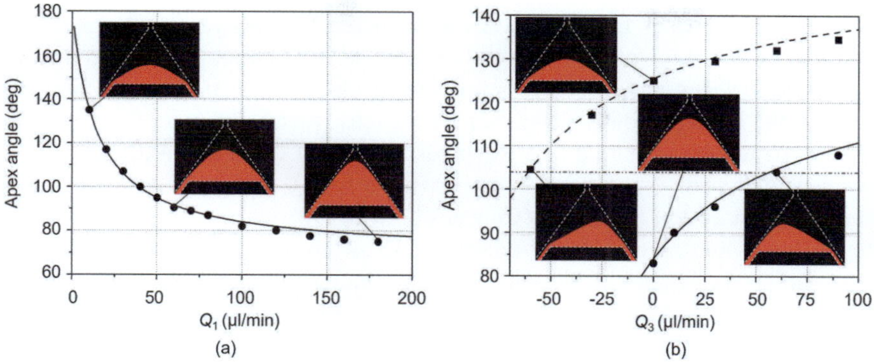

Fig. 3.4 **a** Relationship between the apex angle of the prism and the flow rate Q_1, with Q_2 fixed at 10 μl/min. **b** Relationship between the apex angle of the prism and the flow rate Q_3, with Q_2 fixed at 120 μl/min and Q_1 fixed at 10 μl/min (solid line) and 70 μl/min (dash line). Reproduced from [9] with permission from the Royal Society of Chemistry

Asymmetric prism shapes are achieved with the assistance of the control flow. Figure 3.4b illustrates the prism shape as a function of the control flow rate. The inner flow rate is fixed at 120 μl/min, while the outer flow rates are set at 10 and 70 μl/min, respectively. The control flow rate varies from −60 to 90 μl/min. When both the inner and outer flow rates are fixed, the apex angle of the prism increases with the control flow rate. Unlike the symmetric prism, the vertex of the prism is no longer centered in the chamber; it shifts from the right to the left side as the control flow rate increases, as shown in Fig. 3.4b. The base angle β_1 remains unchanged by the control flow when the inner and outer flow rates are fixed. However, the base angle β_2 varies with the control flow rate, altering the prism shape. When the control flow exits the chamber ($Q_3 < 0$), the volume of flow to the original outlet decreases, reducing the outer flow rate at the chamber's outlet side. According to Eqs. (3.2) and (3.3), the base angle β_2 is larger than β_1. Thus, the vertex of the prism is positioned on the outlet side of the chamber, forming a right-shifting prism. Conversely, when the control flow enters the chamber ($Q_3 > 0$), the outer flow rate increases at the outlet side of the chamber. Consequently, the base angle β_2 becomes smaller than β_1, positioning the vertex of the prism on the inlet side and forming a left-shifting prism. For example, when the outer flow and control flow rates are 10 and 60 μl/min, respectively, the base angles β_1 and β_2 are measured as 48° and 27°, resulting in a left-shifting prism with an apex angle of 105°. Conversely, when the outer flow and control flow rates are 70 and −60 μl/min, the base angles β_1 and β_2 are 27° and 48°, forming a right-shifting prism with an apex angle of 105°. Therefore, prisms with same apex angle but different shapes can be achieved by adjusting the control flow. The tunable optofluidic prism is realized by altering its shape through flow rate control.

To investigate the stability of the prism, experiments were conducted 10 times, with 5 data points measured every 30 s in each run. The standard deviation of the prism's apex angles ranged from 0.66° to 1.08°. However, when the inner flow was replaced with DI water, the standard deviation increased to between 1.11° and 1.95°. This indicates that the prism made with benzyl alcohol is more stable than the one made with DI water. The difference is attributed to the mismatch in Reynolds numbers between the inner and outer flows. For a symmetric prism, the outer flow rate was fixed at 10 μl/min, resulting in a Reynolds number of 0.23. When the inner flow was benzyl alcohol, the Reynolds number varied from 0.64 to 5.73 as the flow rate changed from 20 to 180 μl/min. Conversely, with DI water as the inner flow, the Reynolds number ranged from 14.98 to 67.42 as the flow rate varied from 80 to 360 μl/min. The larger difference in Reynolds numbers between the outer mixture and the inner DI water flow makes the prism shape more unstable when constructed with DI water.

3.3 Light Scanning via Optofluidic Prism

When the optofluidic prism is created within the microchamber, the direction of the light beam changes as it passes through the prism. Assuming a light beam enters the prism horizontally, as illustrated in Fig. 3.5, the total deviation angle, δ, can be determined using Snell's law. The prism structure comprises two surfaces: A'B and A'C. The total deviation angle is given by

$$\delta = (i_1 - i_3) + (i_2 - i_4). \tag{3.5}$$

Here, i_1 and i_3 represent the incident angle and refractive angle at the A'B interface, respectively, while i_4 and i_2 represent the incident angle and refractive angle at the A'C interface. These angles can be calculated using geometric optics principles:

$$\begin{cases} i_1 = 90^\circ - \beta_1, \\ i_2 = \arcsin(n \sin i_4), \\ i_3 = \arcsin(1/n \cos \beta_1), \\ i_4 = 180^\circ - \beta_2 - \beta_1 - i_3, \end{cases} \tag{3.6}$$

where $n = n_1/n_2$, with n_1 and n_2 being the refractive indices of the inner and outer fluids, respectively. For a symmetric prism, using Eqs. (3.5) and (3.6), the total deviation angle can be expressed as

$$\delta = i_1 + \arcsin\left\{\sin(\phi)\sqrt{n^2 - \sin^2(i_1)} - \cos(\phi)\sin(i_1)\right\} - \phi. \tag{3.7}$$

The displacement of the output light beam (at point G) can also be estimated using geometric optics principles. It is important to note that the deviation angle and output position of the light beam are not solely determined by the prism's shape (e.g., as influenced by the base angles β_1 and β_2), but also by the refractive indices of the

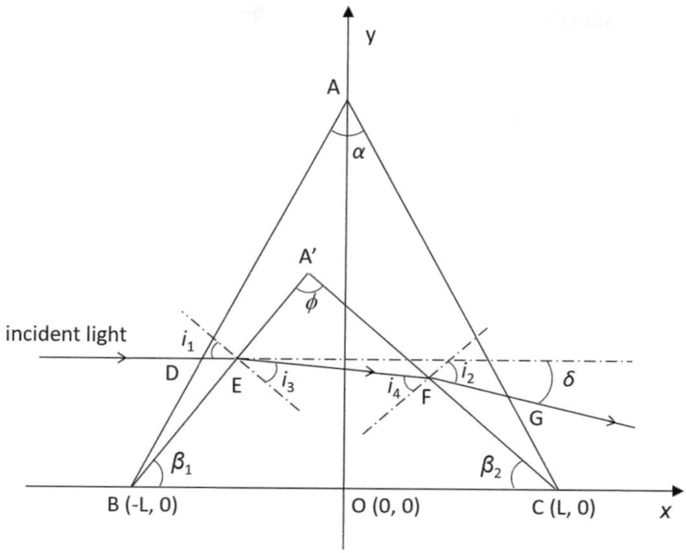

Fig. 3.5 Optical model of the optofluidic prism. Reproduced from [9] with permission from the Royal Society of Chemistry

media involved. In particular, for a liquid prism, the refractive index can be adjusted by altering the inner or outer fluids, leading to spatial shifts in the optical beam paths.

To illustrate the adjustable optical path of the optofluidic prism, beam-tracing experiments are conducted on symmetric prisms. Figure 3.6 displays the changes in the deviation angle of the output beam under various flow conditions. The light beam is horizontally incident on the prism, and the light path in the beam-tracing chamber appears green due to the excited fluorescent dye. A positive (negative) deviation angle indicates the output beam is refracted downward (upward). Both 70° and 90° apex angle chambers are used to examine the influence of chamber shape on the prism shape and the adjustable deviation angle of the output beam. The outer flow, an ethylene glycol solution ($n = 1.412$), is maintained at a flow rate of 10 μl/min. The inner flow, consisting of DI water or benzyl alcohol, varies from 20 to 400 μl/min. Utilizing two types of inner fluids aims to broaden the tuning range of the deviation angle.

When the refractive index of the inner flow exceeds than that of the outer flow (e.g., benzyl alcohol as the inner fluid), the output light beam is refracted downward, as shown in Fig. 3.6a. The relationship between the apex angles of the prisms and the deviation angles of the output beams is plotted against the flow rate of the inner flow, which is benzyl alcohol. As the flow rate of the inner flow varies from 20 to 180 μl/min, the apex angle of the prism in a 70° chamber changes from 116.5° to 78°, and in a 90° chamber, it changes from 128° to 97°. Though the apex angle of the prism continues to decrease with an increasing flow rate beyond 180 μl/min, the rate of decrease slows significantly. Additionally, the apex angle of the prism will not be

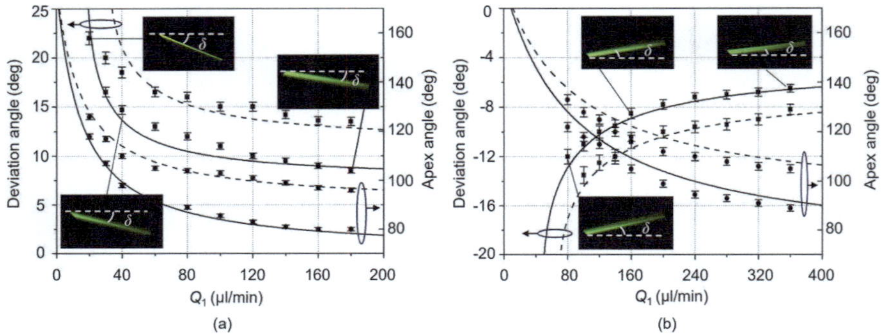

Fig. 3.6 Relationship between the deviation angle, apex angle and the flow rate Q_1 with chambers of 70° (solid line) and 90° (dash line). The inner liquid is **a** benzyl alcohol ($n = 1.540$) and **b** DI water ($n = 1.332$). Reproduced from [9] with permission from the Royal Society of Chemistry

smaller than that of its chamber. As the apex angle of the prism increases, the height of the prism decreases. Considering the fixed position of the fiber, only prisms with an apex angle smaller than 130° are examined. The deviation angles corresponding to the 70° chamber range from 21° to 8.5°, while for the 90° chamber, they range from 22° to 11.5°.

When the refractive index of the inner flow is lower than that of the outer flow (e.g., DI water as the outer fluid), the output light beam refracts upwards, as shown in Fig. 3.6b. The apex angle of a prism made with DI water is larger compared to one made with benzyl alcohol at the same flow rates due to the much lower viscosity of DI water. At an inner flow rate of 200 µl/min, the apex angle of the prism is 103° in the 70° chamber and 116° in the 90° chamber. To extend the tuning range of the deviation angle, the inner flow rate is varied from 40 to 360 µl/min. This results in deviation angles ranging from −12° to −6.5° in the 70° chamber and from −13.5° to −8° in the 90° chamber.

The variation in the deviation angle aligns with the changes in the prism's shape, as expected from geometrical optics: a prism with a larger apex angle results in a greater deviation angle. Additionally, the experimental results support the theoretical prediction that a chamber with a smaller apex angle can achieve a wider tuning range for the prism's apex angle. However, the tuning range of the deviation angle is not significantly expanded in the chamber with a smaller apex angle. The notable differences appear in the working zone. The choice of the chamber shape should be based on the specific requirements for the output light. For the benzyl alcohol and 70° chamber system, the working zone of the deviation angle is 21° to 8.5°. If a deviation angle of 7° is desired, the chamber should be designed with a smaller apex angle (e.g., 50°) or a liquid with a lower refractive index should be selected. Using the existing design and chosen liquids, the deviation angle can be adjusted from −13.5° to 22°. Achieving a larger deviation angle is possible by using other outer and inner liquids with a higher refractive index contrast.

For the symmetric prism, altering the flow rate ratio between the inner and outer flows changes the deviation angle of the output light beam, as well as the position of the output beam (point G in Fig. 3.5). It is challenging to control both the output position and the propagation direction of the light beam simultaneously without modifying the properties of the liquids. This limitation can be overcome with an asymmetric prism design, which offers more precise and flexible control over the optical path. For example, Fig. 3.7a demonstrates that the same deviation angle can be achieved using different combinations of flow rates Q_2 and Q_3. This means one deviation angle corresponds to multiple positions of the output light beam. In a typical experiment, the inner flow rate Q_1 is maintained at 100 μl/min, while the outer flow rate Q_2 and the control flow rate Q_3 are adjusted simultaneously to keep the deviation angle constant at 10°. The relationship between Q_2 and Q_3 is depicted as a dashed line in Fig. 3.7b. The calculated output positions vary with the flow rates and are plotted as a function of Q_2 in a solid line. Several combinations of Q_2 and Q_3 are tested, each producing a refracted light beam with a deviation angle of 10°. As Q_2 increases, the light beams are observed to shift downward parallelly with a deviation angle of 10°. The position of the output light is measured from images captured by a CCD camera and plotted as a function of Q_2. The results indicate a tuning range of approximately 60 μm. A larger tuning range can be achieved by further optimizing the refractive indices and chamber geometries.

To demonstrate the potential for adjusting the alignment of the optical path between two components, an optical fiber is fixed while parallel light scanning is performed. Theoretically, the optical fiber is aligned to receive light with a deviation angle of 10° at the optimal position ($Q_2 = 10.5$ μl/min). In the experiment, parallel light scanning is carried out, and the detected intensity is plotted in terms of Q_2. The maximum light intensity is not at the predicted position but occurs when the asymmetric prism is formed ($Q_2 = 9.5$ μl/min, $Q_3 = 1.7$ μl/min), with the light intensity

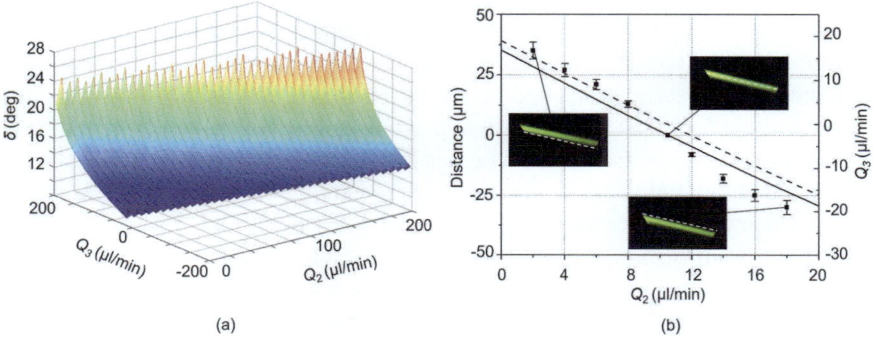

Fig. 3.7 **a** Relationship between the deviation angle and the flow rates Q_2 and Q_3, with Q_1 fixed at 10 μl/min. **b** Light scanning controlled by Q_2 and Q_3, with Q_1 kept at 100 μl/min. The output light position (solid line) and Q_3 (dashed line) are plotted as functions of Q_2. Reproduced from [9] with permission from the Royal Society of Chemistry

increasing by 65.7% compared to the theoretically predicted position. This highlights the importance of optical path alignment in integrated micro-optical circuits and demonstrates the strength of the developed optofluidic prism. The potential applications for the optical-based lab-on-a-chip such as localized or parallel scanning fluorescent excitation can take great advantages over the tunable optofluidic prism.

3.4 Optofluidic Prism Refractometer

In the earlier sections, the output beam was mainly controlled by adjusting the flow rates of the fluid streams in the optofluidic chip. Additionally, the optofluidic prism is sensitive to the refractive indices of the flows, enabling the creation of an on-chip refractometer. The triangular chamber is filled with a single fluid, which means the prism shape remains fixed and the deviation angle of the light beam depends on the fluid's refractive index. A chamber with a 70° apex angle is used here. Since the chip is made of PDMS, the reference refractive index n_2 is 1.412. The tested samples are mixtures of benzyl alcohol and water, with the refractive index (n_1) varying from 1.332 to 1.540 by adjusting the mixing ratio. The deviation angle of the output beam shifts with changes in the refractive index as expressed in Eq. (3.7). The output light position can be calculated accordingly. Because the light beam can be traced in a nearby chamber filled with fluorescent dye (Fig. 3.8), the easily identifiable light beam position can be used to deduce the refractive index of the sample in the prism chamber. Adding markers on the chip can enhance the usability of the on-chip refractometer.

Figure 3.8 demonstrates the variation in the output light position as a function of the refractive index of the tested sample. This overlay image combines experimental results from four different solutions ($n_1 = 1.359$, 1.412, 1.477 and 1.540). As the prism's refractive index increases, the deviation angle also increases, causing the corresponding light beam to shift downward. The experimental results are compared with simulation results, as marked on the beam-tracing chamber in Fig. 3.8. This comparison confirms that the output light position closely aligns with the predictions derived from the theoretical model based on Eqs. (3.5)–(3.7). Figure 3.9a depicts the normalized output light intensity profile shifting with the sample's refractive index. The light beam profile is taken along line FF', as shown in Fig. 3.8. The peak position of the light beam is plotted against the refractive index of the solution. The point $x = 0$ represents the position where the deviation angle of the light beam is zero, indicating the sample's refractive index matches that of PDMS ($n_1 = n_2 = 1.412$). As the refractive index varies from 1.358 to 1.540, the light beam shifts by 320 μm.

Figure 3.9b illustrates the position of the light beam as a function of the refractive index, showing a linear relationship between the two parameters. Using a solution with a standard refractive index, experimental data can calibrate the position of markers in the on-chip refractometer. A CCD detector with a resolution of 0.339 μm/pixel achieves a sensitivity of 2×10^{-4} RIU/pixel. Moreover, the sample volume

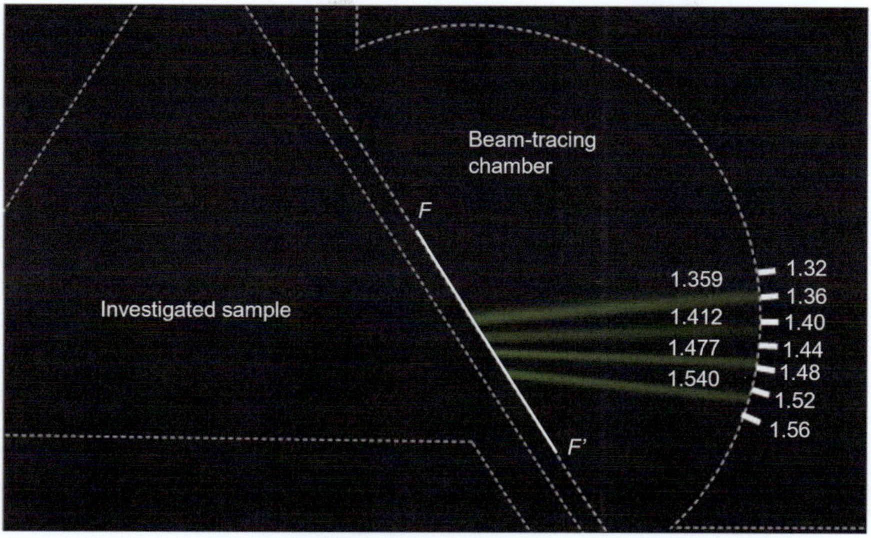

Fig. 3.8 Microscopic imaging of the on-chip refractometer

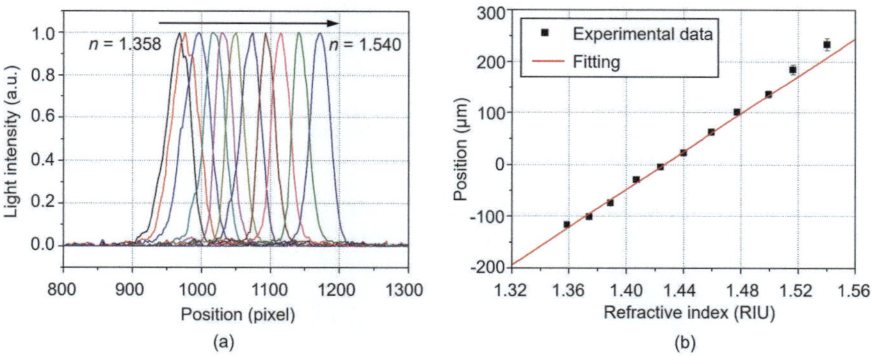

Fig. 3.9 **a** Output light intensity profile shifts with changes in the refractive index. **b** Position of the light beam is plotted as a function of the refractive index

required for analysis is just 0.35 μl. This optofluidic prism-based on-chip refractometer offers a new method for on-chip analysis, requiring only a small sample quantity and being easily integrable into lab-on-a-chip systems.

3.5 Summary

Precise adjustment of the optical path among multiple integrated components is crucial for optical detection in microfluidic devices. For instance, the curvatures of optofluidic microlenses can be fine-tuned to achieve optimal performance in an optical detection system. However, microlenses have limited beam-steering capabilities. In traditional optical systems, prisms are essential for altering the optical path of a collimated light beam or performing light dispersion. Therefore, realizing a tunable optofluidic prism that can precisely adjust the optical path in an integrated on-chip system is one of the requests in the development of lab-on-a-chip systems. Additionally, the high sensitivity of optofluidic prisms in measuring the refractive indices of analytes is advantageous. This chapter introduces a tunable optofluidic prism. A simple theoretical model predicts the apex angle of the prism, the deviation angle, and the position of the output light beam. By precisely controlling the prism shape, parallel light beam scanning is achieved. It is also possible to utilize the optofluidic prism as an on-chip refractometer. The optofluidic prism holds potential for creating on-chip optical circuits and shows promise for optical detection and biochemical analysis.

References

1. Starkey TV (1938) A variable-angle liquid-prism spectrometer. J Sci Instrum 15(9):290
2. Stanislaw B (1993) Prism with a variable refraction angle and refractive index. Phys Educ 28(4):256
3. Yoshihata Y, Takei A, Binh-Khiem N, Kan T, Iwase E, Matsumoto K, Shimoyama I (2009) Micro liquid prism. In: IEEE 22nd international conference on micro electro mechanical system (MEMS)
4. Takei A, Iwase E, Hoshino K, Matsumoto K, Shimoyama I (2007) Angle-tunable liquid wedge prism driven by electrowetting. J Microelectromech Syst 16(6):1537–1542
5. Kou Q, Yesilyurt I, Studer V, Belotti M, Cambril E, Chen Y (2004) On-chip optical components and microfluidic systems. Microelectron Eng 73–4:876–880
6. Llobera A, Wilke R, Buttgenbach S (2004) Poly(dimethylsiloxane) hollow Abbe prism with microlenses for detection based on absorption and refractive index shift. Lab Chip 4(1):24–27
7. Llobera A, Wilke R, Buttgenbach S (2005) Optimization of poly(dimethylsiloxane) hollow prisms for optical sensing. Lab Chip 5(5):506–511
8. Song CL, Nguyen NT, Asundi AK, Tan SH (2010) Tunable micro-optofluidic prism based on liquid-core liquid-cladding configuration. Opt Lett 35(3):327–329
9. Xiong S, Liu AQ, Chin LK, Yang Y (2011) An optofluidic prism tuned by two laminar flows. Lab Chip 11:1864–1869
10. Wu ZG, Nguyen NT (2005) Convective-diffusive transport in parallel lamination micromixers. Microfluid Nanofluid 1(3):208–217
11. Larsen MU, Shapley NC (2007) Stream spreading in multilayer microfluidic flows of suspensions. Anal Chem 79(5):1947–1953

Chapter 4
Optofluidic Hyperbolic Secant Lens

4.1 Introduction

Diffusion is the process where molecules move from an area of high concentration to one of low concentration. It plays a crucial role in microfluidics, being the primary means of liquid mixing in the laminar flow regime. This method of fluid–fluid interaction is vital for creating novel optofluidic components. Gradient index profiles can be generated and adjusted by controlling liquid diffusion within microchannels. Light passing through these index profiles exhibits specific behaviors like focusing and diverging. The classic example of diffusion-controlled optofluidic components is the optofluidic liquid-core liquid-cladding waveguides [1]. In this configuration, high and low refractive index liquids serve as the core and cladding flow streams, respectively, and diffusion between these streams creates a gradient index profile. An optofluidic focusing lens has been demonstrated using this setup [2]. Similarly, large-angle bending can be achieved by reversing the positions of the high and low index liquids, using the low index liquid as the core stream and the high index liquid as the cladding streams [3]. In another example, diffusion in five flow streams is used to create an optical beam splitter, with the splitting ratio tunable through dynamic flow rate control [4]. The splitting ratio is tunable by dynamically controlling the flow rates. Additionally, a wavelength filter is achieved by introducing dye molecules into specific flow streams.

Optofluidic lenses are crucial components for on-chip optical systems, offering distinct advantages such as a tunable refractive index and reconfigurable geometry. Chapter 2 introduced optofluidic microlenses with tunable geometric shapes based on the hydrodynamic focusing effect. Another popular type of optofluidic lens is the Gradient Refractive Index (GRIN) lens, which uses a GRIN to gradually bend light beams over a large area rather than refracting them instantly at an interface. The light propagation in optofluidic GRIN lenses can be calculated using the ray tracing method. The GRIN profile can be generated through various methods, with the most common being controlled liquid diffusion at low Reynolds numbers. In this method,

S. Xiong and A.-Q. Liu, *Optofluidics: Fundamentals and Applications*,
Microfluidics and Nanophotonics: Science and Engineering 4,
https://doi.org/10.1007/978-981-95-6296-1_4

liquids with different refractive indices are pumped into a microchannel at low flow rates. Diffusion between the flow streams creates a GRIN profile, used to focus or split light beams. A diffusion-induced GRIN lens, demonstrated using $CaCl_2$ and DI water, offers a two-dimensional variable focal length within a range of several hundred micrometers by adjusting the fluidic configuration [5]. Another study optimized the index profile to reduce spherical and field curvature aberrations, resulting in a spot size smaller than 10 μm [6]. This configuration requires low flow rates to form the lens, resulting in lower solution consumption compared to liquid–liquid interface-based lenses. However, generating a three-dimensional gradient index is challenging, and thus, reported GRIN lenses are usually two-dimensional, limiting their use in imaging applications. Another method to generate the GRIN involves the temperature-dependent nature of liquids. Several thermal GRIN lenses with various configurations have demonstrated light focusing [7–9]. A thermal-induced GRIN lens has been demonstrated using co-flowing streams with different temperatures [7]. Another thermal-induced GRIN lens has been created by shining laser beams on two chromium strips [9]. The benzyl alcohol solution, heated by the metal strips, forms a two-dimensional gradient, changing the focal length from infinity to a minimum of 1.3 mm. The main issue with this configuration is the limited refractive index contrast due to the low thermo-optical efficiency.

4.2 Optical Aberrations in Optofluidic Lenses

Optofluidic lenses, which can adjust the focal length in two dimensions, are used in integrated sample illumination and optical detection [10, 11]. However, their optical performance can be hindered by optical aberrations such as spherical and field curvature aberrations, limiting their applicability [12, 13]. Here, spherical aberration refers to the phenomenon where incident light beams are focused at multiple spatial locations by the lens, as depicted in Fig. 4.1a. In a conventional optofluidic GRIN lens, light with incident angles θ_1 and θ_2 is focused at two positions, resulting in a spherical aberration of δ. In contrast, an optimized optofluidic lens reduces spherical aberration to zero, as shown in Fig. 4.1c. Reducing spherical aberration in optofluidic lenses can significantly enhance optical performance in on-chip flow cytometry and integrated optical detection. For example, in fluorescent spectroscopy, the limit of detection is improved threefold with a reduced beam waist of 80 μm [10]. In on-chip flow cytometry, a smaller beam waist ensures that a single sample is excited at a time in the detection region, allowing for higher sample concentrations and better performance in terms of signal-to-noise ratio, coefficient of variance (CV), and throughput [11, 14].

Field curvature aberration occurs when multiple light sources placed at different off-axis positions are focused on different focal planes, leading to unwanted signal variations in detection. Traditionally, integrated beam splitters or lens arrays are used to generate multiple light sources, increasing the footprint and complexity of the configuration [15]. In a conventional optofluidic GRIN lens, field curvature aberration

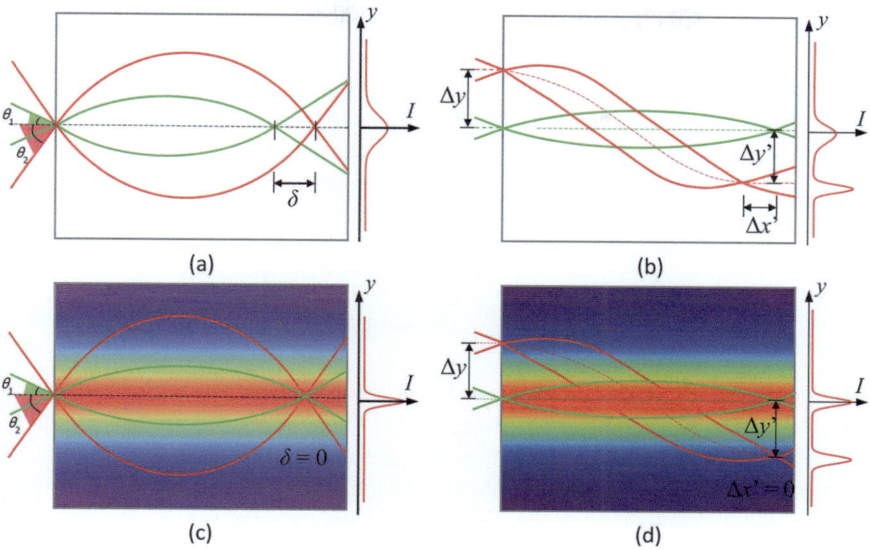

Fig. 4.1 Schematic illustration of **a** spherical and **b** field curvature aberrations in conventional optofluidic GRIN lenses; and the optofluidic lens without **c** spherical and **d** field curvature aberrations. Reproduced from [6] with permission from the Royal Society of Chemistry

causes off-axis light to focus on a focal plane closer to the lens ($\Delta x' \neq 0$, $\Delta y' \neq \Delta y$), resulting in focal points that are not equally sharp for different off-axis positions, as shown in Fig. 4.1b. In contrast, an optofluidic lens with an optimized index profile can focus light on the same focal plane ($\Delta x' = 0$) and at symmetrical positions ($\Delta y' = \Delta y$), as shown in Fig. 4.1d. By suppressing field curvature, a single optofluidic lens can focus multiple light sources on the same focal plane. This makes the single optofluidic lens suitable for multiplexed detection applications such as barcoded particle scanning, particle velocity measurement, flow cytometry, photobleaching quantification, and more.

4.3 Microfluidic Gradient Generator and Index Profile

To minimize spherical and field curvature aberrations, the optofluidic lens is designed using the aberration-free Maxwell's fisheye lens, which was first developed by Maxwell in 1854. The index profile is given by

$$n(r) = \frac{n_0}{1 + \left(r/R\right)^2}, \tag{4.1}$$

where n_0 is the index at the center point, R is the lens radius, and r is the distance to the lens center. Maxwell's fisheye lens is renowned for its aberration-free nature and excellent optical performance due to its spherically symmetric gradient-index design. However, in microfluidics, it is challenging to generate a precisely symmetric gradient index in a circular geometry. Therefore, the circular Maxwell's fisheye lens with a two-dimensional gradient is transformed into a rectangular lens with a one-dimensional gradient using coordinate transformations. The index with a hyperbolic secant profile is expressed as

$$n(y) = \frac{n_0}{2}\operatorname{sech}(\alpha y), \tag{4.2}$$

where n_0 is the highest index value and α is a constant related to the lens geometry. This flattened version of Maxwell's fisheye lens is termed as the optofluidic hyperbolic secant (HS) lens. Figure 4.2a, b depict light propagation in both the Maxwell's fisheye lens and the optofluidic HS lens. The grey coordinate lines illustrate the spatial relationship of the index profiles between the two lenses, while the black lines represent the spatial relationship when a light source is focused by the lenses. Theoretically, both lenses are equivalent and should possess the same optical functionalities.

The ideal index profile for the optofluidic lens is a hyperbolic secant profile ranging from 0 to $n_0/2$, based on Eq. (4.2). However, liquid mediums typically have indices between 1.3 and 1.7. Consequently, the top and bottom regions of the rectangular lens (white region) must be discarded, as shown in Fig. 4.3a. In terms of the spatial

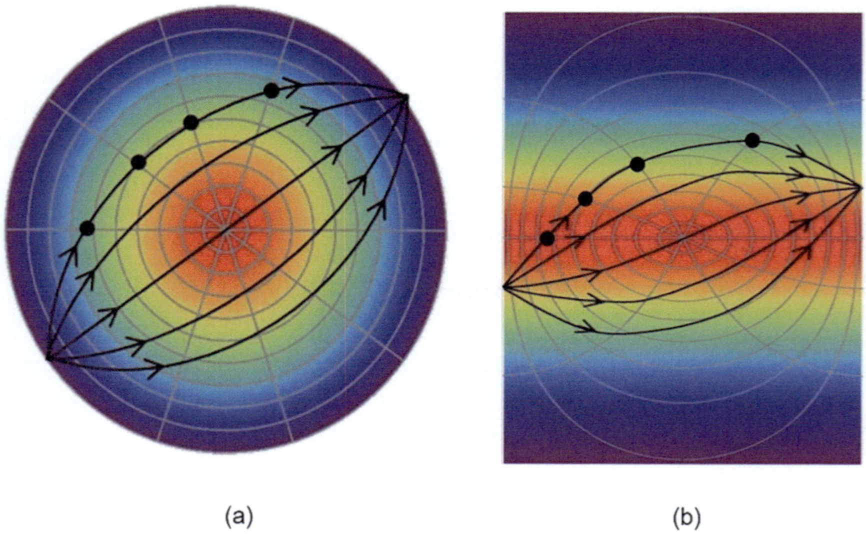

Fig. 4.2 Schematic illustration of light propagation in **a** Maxwell's fisheye lens and **b** optofluidic HS lens. Reproduced from [6] with permission from the Royal Society of Chemistry

relationship between the two coordinate systems, this discarded region corresponds to the pole regions in Maxwell's fisheye lens, as depicted in Fig. 4.3b. When the incident angle of the light beams is large enough to reach these discarded regions, the light can no longer be focused, as illustrated in Fig. 4.3c, d. Therefore, the index approximation reduces the maximum divergence angle. The spatial relationship suggests that a higher index contrast is preferred to minimize the discarded regions and thus achieve a larger maximum divergence angle.

To create an optofluidic HS lens, three flow streams with different concentrations are injected into the microfluidic network, where they undergo repeated splitting,

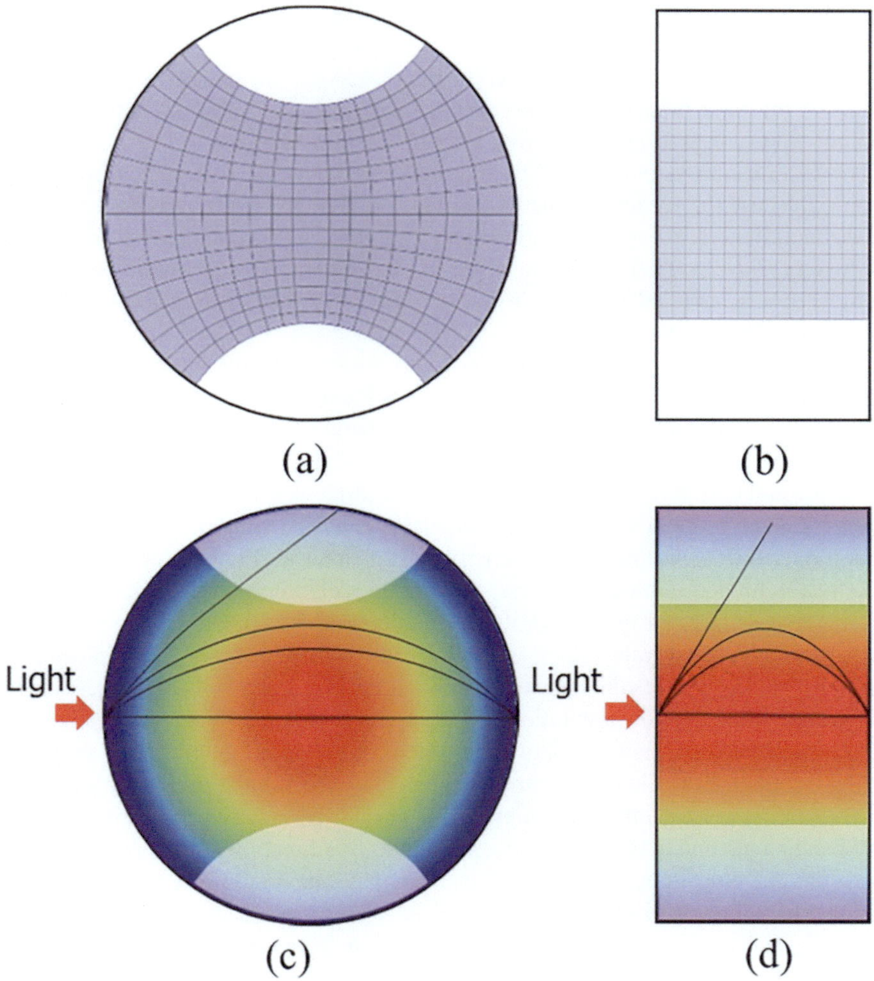

Fig. 4.3 The spatial relationship between **a** the circular fisheye lens and **b** the optofluidic lens. **c**, **d** Light propagation in both lenses with index approximation. Reproduced from [6] with permission from the Royal Society of Chemistry

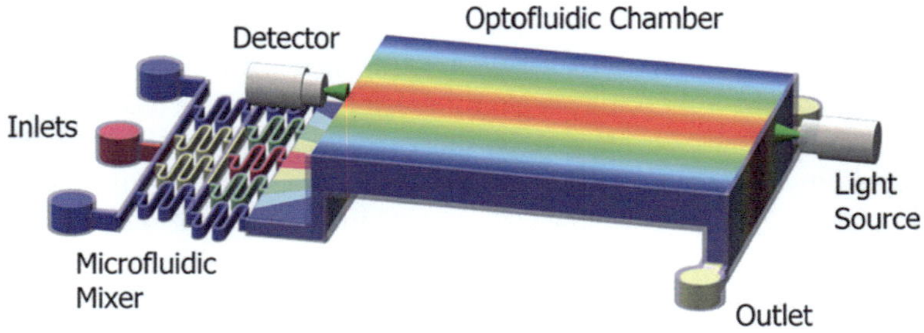

Fig. 4.4 Schematic of the chip design for forming an optofluidic HS lens. Reproduced from [6] with permission from the Royal Society of Chemistry

mixing, and recombination, as shown in Fig. 4.4. After six stages of micromixers, each of the nine branch flows has distinct concentration values. These different branch flows are then combined into a single channel to create the desired profile. The splitting ratio is determined by the pressure above and below each branching point. Because the vertical channel's resistance is negligible compared to that of the horizontal mixers, the splitting ratio can be simplified to the quantitative proportion of mixers above and below the branching point. To generate a parabolic profile, the middle inlet is supplied with a high-index liquid, while the top and bottom inlets are supplied with DI water.

As illustrated in Fig. 4.5, the refractive index profile of the optofluidic lens demonstrates that the flow streams display discrete refractive index values in the branch channels, which gradually mix in the mixing chamber to create a smooth profile. Figure 4.6a presents the refractive index profiles at the position x_1 under different cladding and core flow rates. The index profiles align well with the refractive index formula given by Eq. (4.2). The edge mismatch is attributed to the slower flow rate and longer diffusion time near the microchannel sidewall. Figure 4.6b summarizes the maximum and minimum indices, showing the relationship between the ratio of flow rates and the range of index profiles. It depicts data for a cladding flow of 1.0 μL/min (red) and 10 μL/min (blue), with the core-to-cladding flow rate ratio ranging from 0.2 to 4.0. The results indicate that the index profile exhibits the largest index difference when the flow rate ratio is 2.25 and the cladding flow rate is 10 μL/min.

The spherical and field curvature aberrations of the optofluidic lens are analyzed using ray tracing simulation. Figure 4.7a, b display the simulation results when the light source is positioned at the center and 30 μm above the center. Figure 4.7c shows the focal length and maximum divergence angles as a function of the flow rate ratio, with the cladding flow rate set at 10 μL/min. The maximum divergence angle is defined as the angle at which light is focused with spherical aberration (δ) of less than 15 μm. The data indicates that the focal length decreases sharply to a minimum when the flow rate ratio is below 1.5 and then gradually increases as the ratio rises. Conversely, the divergence angle peaks at 35° when the flow rate ratio is 2.25. Referring to Fig. 4.6b, it can be inferred that increasing the refractive index contrast results in a decrease in focal length and an increase in divergence angle. Therefore, a higher

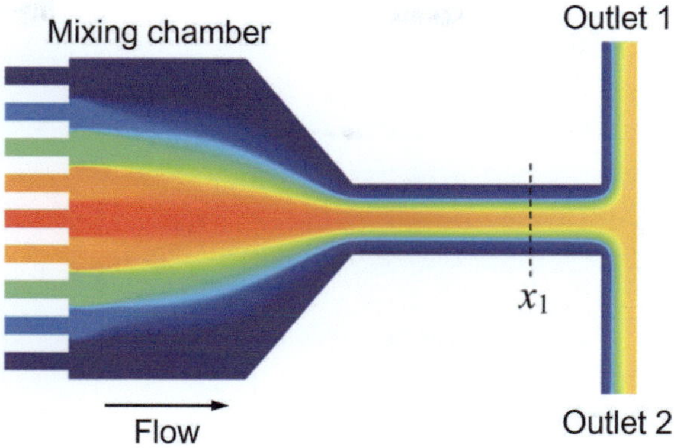

Fig. 4.5 Numerical simulation of the index profile in the optofluidic lens. Reproduced from [6] with permission from the Royal Society of Chemistry

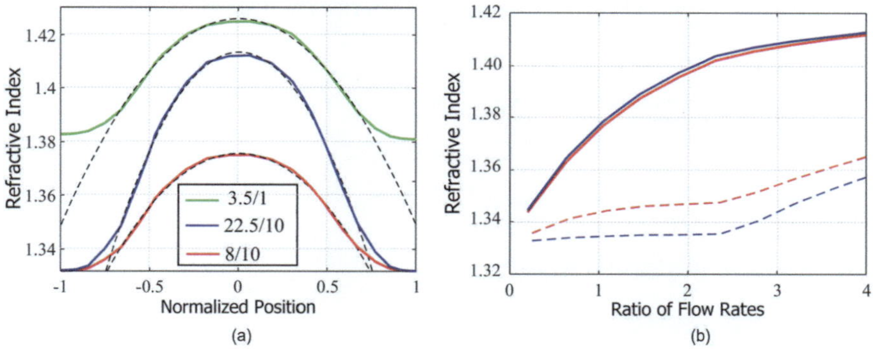

Fig. 4.6 **a** Index profile of the optofluidic lens at varying core/cladding flow rates (solid lines: simulated profiles, dashed lines: ideal fitted profiles). **b** Maximum (solid lines) and minimum (dashed lines) refractive indices at x_1 as a function of the flow rate ratio, with cladding flow rates of 1.0 μL/min (red) and 10 μL/min (blue), respectively. Reproduced from [6] with permission from the Royal Society of Chemistry

index contrast is preferred to better suppress optical aberrations. The maximum divergence angle can be further increased by employing more advanced and precise methods to reduce the refractive index mismatch at the microchannel edges.

Figure 4.7c also shows that the focal length and divergence angle are correlated at a given ratio of flow rates, making it impossible to change the maximum divergence angle and focal length independently. However, this can be achieved by altering the channel width. Figure 4.8 illustrates the simulated focal length as a function of channel width when the flow rate ratio is 2.25, clearly showing that the focal length is linearly proportional to channel widths.

Ray tracing simulation of light beams with different divergence angles shows that the focal length of the outer rays is longer than that of the inner rays in the optofluidic GRIN lens. In contrast, in the optofluidic HS lens, they are focused at the same position on the optical axis. Similarly, ray tracing simulation of light beams with different off-axis positions indicates that when light beams are placed above

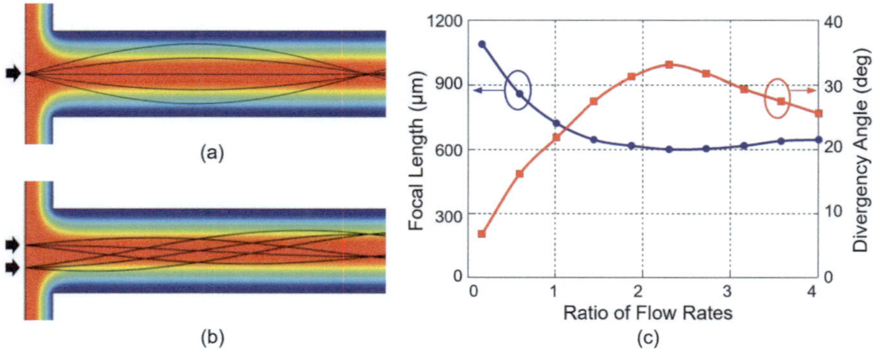

Fig. 4.7 Ray-tracing simulation results when the light source is positioned **a** at the center and **b** 30 μm offset from the center. **c** Focal length and maximum divergence angles as functions of the flow rate ratio. Reproduced from [6] with permission from the Royal Society of Chemistry

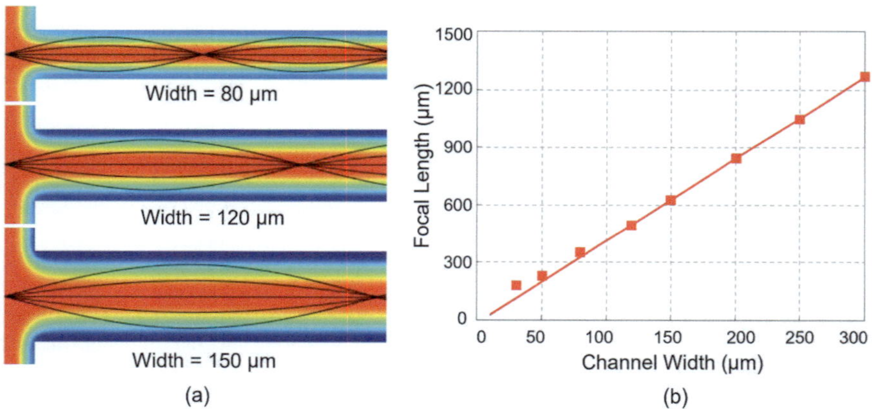

Fig. 4.8 **a** Ray tracing simulation of light beams with different channel widths. **b** Simulated focal length as a function of channel widths with the core/cladding ratio of 2.25. Reproduced from [6] with permission from the Royal Society of Chemistry

the center, the outer and inner rays focus on multiple points in the optofluidic GRIN lens, while they converge to a single point in the optofluidic HS lens.

Based on the ray tracing results, the spherical aberration of the optofluidic GRIN lens and the optofluidic HS lens is compared as a function of the divergence angle, as shown in Fig. 4.9a. The data shows that the spherical aberration δ of both lenses remains under 15 μm when the divergence angle is below 20°. However, as the divergence angle increases, the spherical aberration of the optofluidic GRIN lens rises almost exponentially, while that of the optofluidic HS lens increases more gradually. The results indicate that the optofluidic HS lens extends the maximum divergence angle to 35° for the same δ of 15 μm, making it 1.75 times better than the optofluidic GRIN lens. Similarly, Fig. 4.9b compares the field curvature aberration of

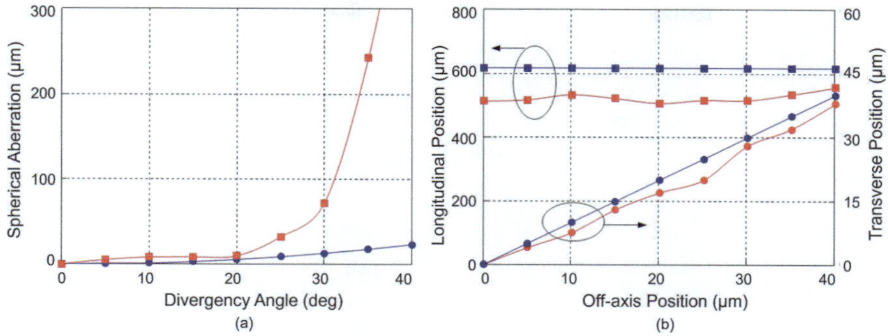

Fig. 4.9 **a** Simulated spherical aberration as a function of divergence angle. **b** Focal position as a function of light source (red: conventional optofluidic GRIN lens, blue: optofluidic HS lens). Reproduced from [6] with permission from the Royal Society of Chemistry

the two lenses based on the ray tracing results. It demonstrates that in the longitudinal direction, the focal positions of the optofluidic GRIN lens shift relative to the off-axis position of the source light, while those of the optofluidic HS lens are on the same focal plane. In the transverse direction, the focal positions of the optofluidic GRIN lens are generally closer to the optical axis, while those of the optofluidic HS lens correspond to the off-axis positions.

4.4 Performance Evaluation of Optofluidic Hyperbolic Secant Lens

The optofluidic chip is fabricated using PDMS with a soft lithography technique. As previously discussed, a higher index contrast is preferred to better suppress optical aberrations. Common liquid mediums typically have refractive indices ranging from 1.3 to 1.7. For this experiment, ethylene glycol and DI water are chosen due to their relatively high index contrast and easy accessibility. Rhodamine 6G is added to both ethylene glycol and DI water to visualize light propagation in the optofluidic HS lens.

The optofluidic chip is placed on an inverted microscope, and the light propagation image is captured using a digital CCD camera. Liquids are pumped into the chip with a high-precision syringe pump. In the experiments to measure the spherical and field curvature aberrations, a single-mode fiber couples the laser light ($\lambda = 514.5$ nm) into the optofluidic chip. The light intensity images are analyzed by intensity profiling using ImageJ.

To validate the microfluidic gradient generator, the index profile in both the mixing chamber and the optofluidic chamber is measured. During the measurement, fluorescent dye is added to ethylene glycol. The refractive index has a linear relationship with the captured pixel intensity. Figure 4.10a shows that the flow streams

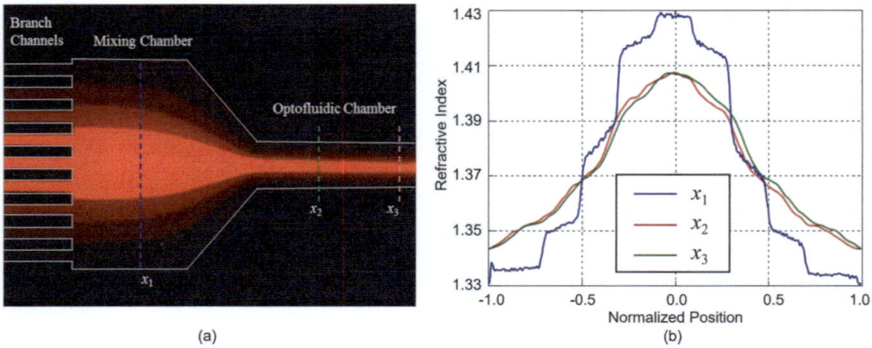

Fig. 4.10 **a** Fluorescent image showing the index profile at various positions. **b** Measured refractive indices at x_1,x_2 and x_3, respectively. Reproduced from [6] with permission from the Royal Society of Chemistry

have discrete refractive index values in the branch channels, which evolve into a continuous hyperbolic secant profile in the mixing chamber.

The measured index profiles at x_1, x_2, and x_3 are plotted in Fig. 4.10b. The blue line indicates that the index profile at x_1 shows nine distinct steps, caused by insufficient mixing of the branch flows. The highest refractive index is 1.428, slightly lower than that of pure ethylene glycol, while the lowest refractive index, found near the sidewalls, is close to that of DI water. Due to the relatively low flow rate in the mixing chamber, the liquids have sufficient diffusion time to transition from a step-index profile to a smooth one. The red and green lines represent the index profiles at x_2 and x_3, respectively, showing that the index profile smooths out as expected.. The highest refractive index decreases to 1.407, and the lowest refractive index increases to 1.343 due to diffusion in the mixing chamber. The results also indicate that the index profiles at x_1 and x_2 are nearly identical, attributed to the high flow rate and short diffusion time in the optofluidic chamber.

To assess the spherical aberration of the optofluidic HS lens, light propagation images are analyzed at different divergence angles. The divergence angle of the light beam from the single-mode fiber is 12°. This angle is increased to 15°, 20°, 25°, 30°, 35°, 40° and 45° using a series of PDMS diverging lenses. These lenses diverge light through the air-PDMS interface at the end of the fiber facet. They are aligned with the center of the optofluidic chamber and positioned 20 μm from the chamber sidewall.

Figure 4.11a–d display light propagation images at divergence angles of 12°, 20°, 30°, and 45°, with cladding and core flow rates of 10 μL/min and 22.5 μL/min, respectively. The red arrows indicate the focal positions. The results demonstrate that the optofluidic HS lens focuses the light beams at the same position for divergence angles below 30°. However, as the divergence angle increases to 45°, a portion of the light beams escapes the optofluidic HS lens and is scattered by the microchannel sidewall, as shown in Fig. 4.11d.

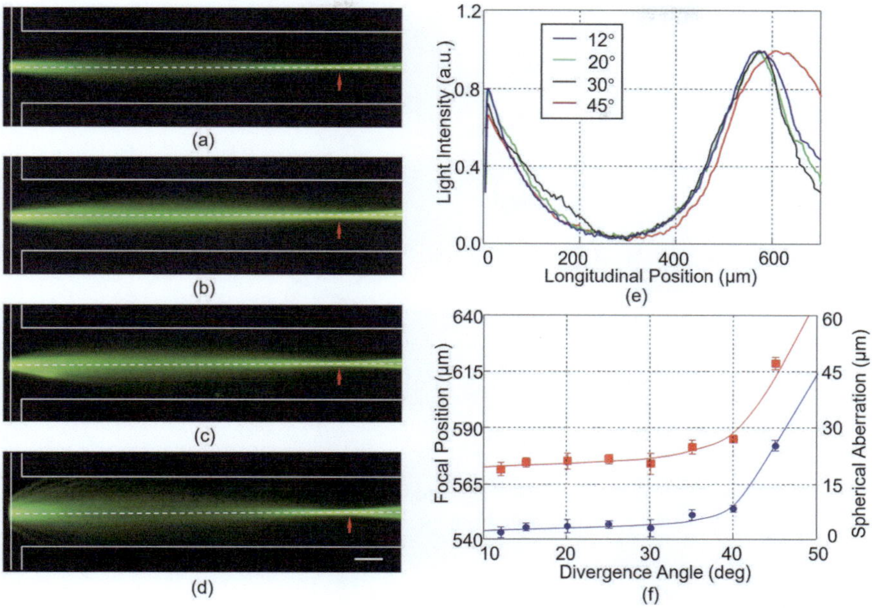

Fig. 4.11 Light propagation images at divergence angles of **a** 12°, **b** 20°, **c** 30° and **d** 45° (scale bar: 50 μm). **e** Light intensity profile along the central line of the microchannel. **f** Focal position (red square) and sperical aberration (blue sphere) as functions of divergence angle. Reproduced from [6] with permission from the Royal Society of Chemistry

Meanwhile, the focal length is slightly longer for divergence angles of 12°, 20°, and 30°. Figure 4.11e plots the light intensities along the central line of the microchannel. The focal length increases slightly from 571.6 to 582.8 μm as the divergence angle increases from 12° to 35°, and significantly to 618.5 μm when the divergence angle reaches 45°. This indicates a considerable increase in spherical aberration at a divergence angle of 45°. In contrast, the beam waists remain identical at divergence angles of 12°, 20°, and 30°, but become much wider at 45°. Figure 4.11f shows the longitudinal focal position and spherical aberration as functions of the divergence angle. Each data point represents the average value of three measured focal positions.

The spherical aberration of the optofluidic lens can be estimated by measuring the paraxial focus (green dot) and the best focal point (black spot) as shown in Fig. 4.12. The focal length of the paraxial rays is estimated as the y-intersection of the focal position curve in Fig. 4.11. The focal length of the best focal point is determined by plotting the intensity profile along the optical axis and identifying the position with the maximum intensity. The distance between the paraxial focus and the best focal points, Δl, is three-quarters of the spherical aberration, δ. At a divergence angle of 35°, the focal lengths of the paraxial rays and the best focal point are estimated to be 568.2 μm and 582.8 μm, respectively. Therefore, the spherical aberration δ of the optofluidic HS lens is 19.5 μm at 35°. The transverse intensity profiles at the

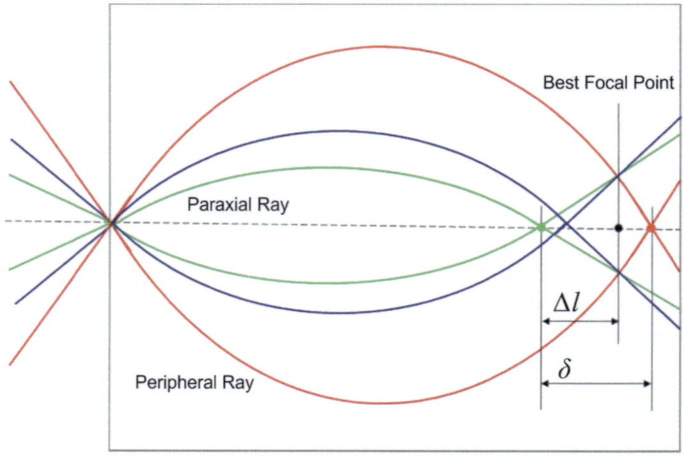

Fig. 4.12 Schematic illustration of spherical aberration measurement. Reproduced from [6] with permission from the Royal Society of Chemistry

best focal point are fitted with a Gaussian function to determine the full width at half maximum (FWHM), which are 6.4, 7.8, 6.3, and 10.6 μm at divergence angles of 12°, 20°, 30°, and 45°, respectively. These results indicate that the optofluidic lens with an HS profile has superior focusing power compared to conventional optofluidic lenses, which typically have a spot size larger than 20 μm [5, 12].

Light propagation versus off-axis positions

To study the field curvature aberration of the optofluidic HS lens, the focal positions in both the longitudinal and transverse directions are examined when the light source is placed at various off-axis positions. The light propagation images for the off-axis positions of 0, 15, 30 and 45 μm are shown in Fig. 4.13a–d. These images indicate that the light sources are focused on the same focal plane in the longitudinal direction. In the transverse direction, the distances between the source and the focal point are 0, 29.8, 58.4, and 84.3 μm at the respective off-axis positions. This result suggests that field curvature is smaller when the light source is closer to the optical axis, with the largest variance ($|\Delta y - \Delta y'|$) being 5.7 μm at an off-axis position of 45 μm.

Figure 4.13e displays the normalized light intensity along the dashed lines shown in Fig. 4.13a–d. By fitting the intensity profiles with Gaussian functions, the FWHM of the focal points is measured to be 8.7 μm, 8.8 μm, 7.8 μm, and 7.6 μm, respectively. The FWHM is nearly the same for off-axis positions of 0 and 15 μm, while it reduces slightly at off-axis positions of 30 and 45 μm. This reduction occurs because, at larger off-axis positions, a portion of the light beams escapes the optofluidic HS lens, resulting in a smaller beam waist at the focal position. The FWHM variance can be further minimized by using liquids with higher index contrast, thereby better suppressing field curvature aberration.

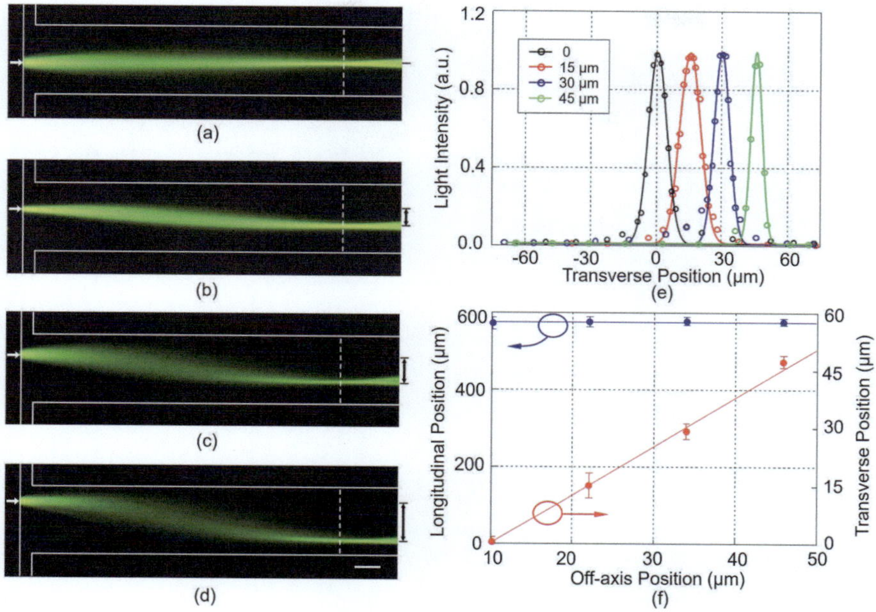

Fig. 4.13 Light propagation images when the light source is placed with the off-axis position of **a** 0, **b** 15, **c** 30 and **d** 45 μm (scale bar: 50 μm). **e** Light intensities on the focal plane. **f** Focal position as a function of off-axis position. Reproduced from [6] with permission from the Royal Society of Chemistry

Figure 4.13f summarizes the transverse and longitudinal focal positions as a function of the off-axis position. Each data point represents the average value of three measurements. The results clearly indicate that the light beams are focused on the focal plane at 572.3 ± 3.4 μm in the longitudinal direction and at opposite positions with a variation of 5.7 μm in the transverse direction.

Light focusing for multiple light sources

The optofluidic HS lens can focus multiple light sources with reduced optical aberrations. Figure 4.14a schematically illustrates its application in multiplexed detection. The lens can uniformly focus an array of light sources, minimizing the difference in illumination intensity at the detection region. Additionally, because of the low dispersion of ethylene glycol and water, the lens can focus multicolor light beams with chromatic aberration of less than 5 μm, according to simulation results. When a barcoded particle flows through the detection region, it will be excited by the focused laser beams of specified wavelengths.

Figure 4.14b displays the light propagation image with two light sources: an argon laser ($\lambda = 514.5$ nm) and a solid-state laser ($\lambda = 532$ nm) used in the experiment. The two light sources are coupled into the optofluidic chamber through two closely placed bare fibers. The distance between the two fiber cores is 125 μm, and the optofluidic chamber has a width of 200 μm. It clearly shows that the two laser sources enter

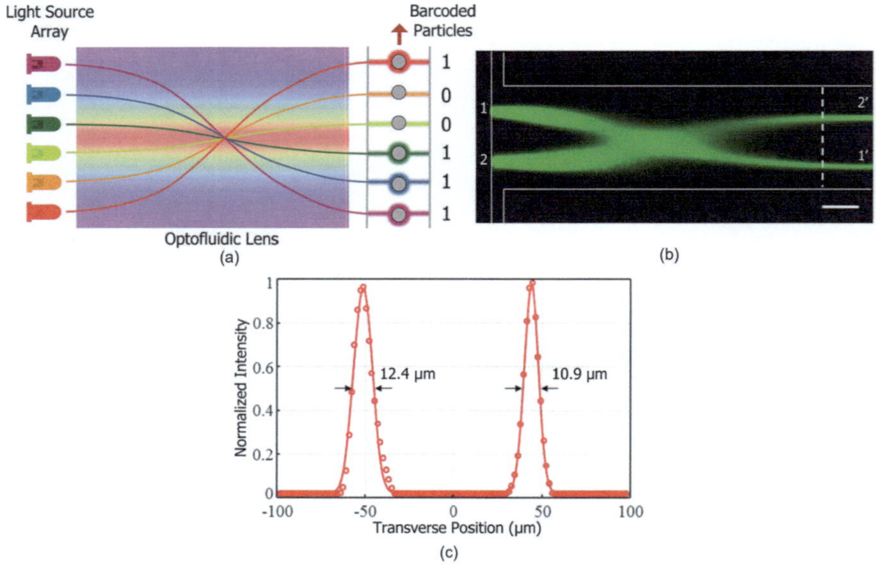

Fig. 4.14 **a** Schematic illustration of the potential application in multiplexed detection. **b** Light propagation image with two light sources (scale bar: 100 μm, vertical direction is magnified by 1.5-fold). **c** FWHM of the two focal points. Reproduced from [6] with permission from the Royal Society of Chemistry

the optofluidic HS lens at positions 1 and 2 from the left sidewall. The laser beams then gradually bend and focus on the opposite transverse positions. Since the two laser beams are incoherent, there is no interference between them in the optofluidic HS lens. Figure 4.14c plots the normalized light intensity along the dashed line in Fig. 4.14b. The FWHMs of sources A and B are measured to be 10.9 μm and 12.4 μm, respectively. The slight FWHM difference is likely caused by the beam profile mismatch between the two laser sources. Using two lasers with similar beam profiles is expected to reduce the beam waist variation in practical applications.

4.5 Summary

Optimizing the index profile can enhance the performance of optofluidic lenses. Previous optofluidic lenses primarily focused on demonstrating tunable light focusing, often overlooking optical performance, which is crucial for certain applications. This chapter discusses the impact of spherical and field curvature aberrations on optical performance and derives the ideal index profile using the coordinate transformation of Maxwell's fisheye lens. It provides an example of designing an optofluidic component from its solid counterpart. Traditionally, optofluidic components were directly adapted from solid ones, retaining the geometry and index profile in the microfluidic implementation. However, the geometry and index profile of many solid optical devices, especially those with advanced functionalities, are challenging to realize in microfluidics. Coordinate transformation may simplify the configuration of such optical devices for microfluidic implementation. This design approach is expected to inspire the development of more integrated optofluidic devices with complex functionalities.

References

1. Yang Y, Liu AQ, Chin LK, Zhang XM, Tsai DP, Lin CL et al (2012) Optofluidic waveguide as a transformation optics device for lightwave bending and manipulation. Nat Commun 3:7
2. Huang H, Mao X, Lin SC, Kiraly B, Huang Y, Huang TJ (2010) Tunable two-dimensional liquid gradient refractive index (L-GRIN) lens for variable light focusing. Lab Chip 10:2387–2393
3. Yang Y, Chin LK, Tsai JM, Tsai DP, Zheludev NI, Liu AQ (2012) Transformation optofluidics for large-angle light bending and tuning. Lab Chip 12:3785–3790
4. Wolfe DB, Vezenov DV, Mayers BT, Whitesides GM, Conroy RS, Prentiss MG (2005) Diffusion-controlled optical elements for optofluidics. Appl Phys Lett 87:181105
5. Mao X, Lin SC, Lapsley MI, Shi J, Juluri BK, Huang TJ (2009) Tunable Liquid Gradient Refractive Index (L-GRIN) lens with two degrees of freedom. Lab Chip 9:2050–2058
6. Zhao HT, Yang Y, Chin LK, Chen HF, Zhu WM, Zhang JB et al (2016) Optofluidic lens with low spherical and low field curvature aberrations. Lab Chip 16:1617–1624
7. Tang SKY, Mayers BT, Vezenov DV, Whitesides GM (2006) Optical waveguiding using thermal gradients across homogeneous liquids in microfluidic channels. Appl Phys Lett 88:061112
8. Seta N, Mawatari K, Kitamori T (2009) Individual nanoparticle detection in liquids by thermal lens microscopy and improvement of detection efficiency using a 1-μm microfluidic channel. Anal Sci 25:275–278
9. Chen Q, Jian A, Li Z, Zhang X (2016) Optofluidic tunable lenses using laser-induced thermal gradient. Lab Chip 16:104–111
10. Camou S, Fujita H, Fujii T (2003) PDMS 2D optical lens integrated with microfluidic channels: principle and characterization. Lab Chip 3:40–45
11. Rosenauer M, Vellekoop MJ (2010) Characterization of a microflow cytometer with an integrated three-dimensional optofluidic lens system. Biomicrofluidics 4:043005
12. Tang SK, Stan CA, Whitesides GM (2008) Dynamically reconfigurable liquid-core liquid-cladding lens in a microfluidic channel. Lab Chip 8:395–401
13. Geary JM (2002) Introduction to lens design: with practical ZEMAX examples. Willmann-Bell, Inc.
14. Watts BR, Zhang Z, Xu C, Cao X, Lin M (2012) Integration of optical components on-chip for scattering and fluorescence detection in an optofluidic device. Biomed Opt Express 3:2784–2793
15. Mogensen KB, Kwok YC, Eijkel JCT, Petersen NJ, Manz A, Kutter JP (2003) A microfluidic device with an integrated waveguide beam splitter for velocity measurements of flowing particles by Fourier transformation. Anal Chem 75:4931–4936

Chapter 5
Optofluidic Droplet Grating

5.1 Introduction

In nature, chemical and biological operations occur in micro-sized spaces like cells and their organelles. Microfluidics enables the formation of femto- to picoliter-sized droplets, allowing for the compartmentalization and mimicry of reactions and molecular processes within individual droplets [1]. Immiscible fluids separated by interfaces, along with the interfaces themselves, provide ideal environments for studying microreactions and interfacial phenomena [2–5]. Microdroplets serve as almost perfect chemical reactors, characterized by rapid thermal transfer, efficient mixing, narrow residence times, and the absence of hydrodynamic dispersion. Manipulating droplets composed of water, buffers, and biological fluids in both air and oil has been demonstrated [6, 7]. Microdroplets have proven to be powerful tools in biological research, offering potential for studying in vitro evolution of proteins and RNAs, cell-free cloning and sequencing, genetics, genomics, and proteomics [8, 9].

Beyond their promising characteristics in biochemical systems, microdroplets have also garnered attention for their potential applications as optical components. Their unique features, such as uniformity, periodicity, and self-assembly, make them particularly suitable for optical gratings [10, 11]. For instance, uniform droplets dispersed into a microfluidic channel can form a long-period grating [12, 13]. Two fibers aligned at the ends of the microchannel enable the liquid grating to couple propagating light from core mode to cladding modes, creating a series of attenuation bands at discrete wavelengths in the transmission spectrum. Additionally, the periodic structure of multiphase droplets in a microchannel can act as a diffraction grating, with no fiber coupling required when the incident light shines from above [14]. This chapter explores the use of microfluidic droplets in diffraction gratings and long-period gratings in detail.

S. Xiong and A.-Q. Liu, *Optofluidics: Fundamentals and Applications*,
Microfluidics and Nanophotonics: Science and Engineering 4,
https://doi.org/10.1007/978-981-95-6296-1_5

5.2 Formation of Droplet Grating

Methods for droplet generation

As interest in droplet-based microfluidics increases, more technologies are emerging for the generation and control of droplets. While microdroplet flows driven by surface and interfacial tension gradients have been studied for over a century, it is only in recent years that the microfluidics community has rekindled its interest in microdroplets and their applications.

Microfluidic systems have been employed to produce multiphase fluids in various formats. A phase refers to a distinct state of aggregation of matter, characterized by identical chemical composition and physical state, separated from other material by phase boundaries. Multiphase fluids contain at least two different fluids within a system, each with different chemical compositions. Compared to single-phase flow, multiphase microfluidic flow is more complex, even in simple configurations. Although microdroplets can be formed in microchannels under static conditions [15], utilizing continuous or pseudo-continuous flows has proven to be the most versatile method for high-throughput droplet generation. One major advantage of microfluidic systems is their ability to significantly reduce droplet size distributions compared to bulk methods. Additionally, droplet production can be parallelized to boost throughput and virtually eliminate cross-contamination between adjacent microdroplets.

Microdroplet generation within microfluidic channels can be initiated using electric fields [16–18]. Emerging technologies have integrated electrodes into microdevices, allowing for electrical control over droplet formation. Dielectrophoresis (DEP) and electrowetting on dielectric (EWOD) are widely used techniques. DEP-driven droplet formation relies on the phenomenon where polarized fluids are attracted to areas of higher electric field intensity. EWOD-based generation relies on the observation that an electric field can alter the interfacial energy between a fluid and its contact surface. Besides electric fields, other methods such as micro-injectors and needles are also used for droplet generation [19, 20]. Simple flexible tubing and narrow-gauge needles can replace traditional elastomeric and hard material microfluidic devices.

Microfluidic systems that exploit flow instabilities between immiscible fluids to generate suspended droplets are of particular interest. Droplets can spontaneously form when laminar streams of aqueous reagents are injected into an immiscible carrier fluid. The two most common methods for generating droplets in this way are flow focusing and T-junction geometries. Numerous focused reviews offer detailed insights into droplet generation and flow dynamics [21–23]. In the flow focusing configuration, the dispersed and continuous phases are forced through a narrow region in a microfluidic channel. This design employs symmetric shearing by the continuous phase on the dispersed phase, enabling more controlled and stable droplet generation. The continuous phase applies pressure and viscous stresses that drive the dispersed phase into a narrow strand, which then breaks inside or downstream of the orifice to form droplets. For example, water-in-oil droplets break off from the aqueous fluid stream when the oil shear stress overcomes the interfacial tension that keeps the droplets attached to the aqueous neck. Droplet size and formation rates depend on the combinations of aqueous and oil-phase flow rates, nozzle geometry, respective dynamic viscosities, densities, and interfacial tension between phases.

In the T-junction configuration, the inlet channel carrying the dispersed phase intersects the main channel containing the continuous phase perpendicularly. The two phases form an interface at the junction, and as fluid flow continues, the tip of the dispersed phase enters the main channel. Shear forces from the continuous phase and the resulting pressure gradient cause the head of the dispersed phase to elongate, thinning the neck and eventually breaking the stream into a droplet. For T-junction nozzles at low oil flow rates, the oil flow "pinches" droplets when the aqueous stream blocks the continuous flow channel, creating pressure that leads to droplet generation. In T-junctions, slightly different breakup mechanisms result in dripping/jetting transitions. The aqueous fluid interface moves into the continuous fluid stream, creating a blockage and causing upstream pressure to build until the pressure force overcomes the interfacial tension, "pinching" the droplet off the mainstream. Similar to the dripping regime for flow-focusing and co-flowing nozzles, monodisperse droplets form as the oil shear pulls droplets from the main aqueous stream. Droplet size and formation rates can be adjusted by altering fluid flow rates, channel widths, or the relative viscosity between the two phases.

To maintain steady droplet generation in flow focusing or T-junction setups, aqueous and oil flow rates must be kept sufficiently low to prevent inertial and shear-based jetting. This requirement imposes a strict limitation on the achievable flow rates and droplet generation frequencies for a given nozzle. Therefore, to increase throughput, one must add parallel channels rather than simply increasing the flow rates.

Droplet array formation in T-junction

A T-junction is a common structure for creating droplet arrays in an optofluidic grating chip. In the upcoming examples, water-in-oil droplets are generated in a T-junction, with detailed discussions on microchannel properties, liquid selection, droplet formation processes, and size control, among other topics.

Flow in microchannels can be driven not only by external pressure or forces generated by electrical, thermal, photonic, or acoustic fields but also by internal forces determined by fluid and channel properties, such as interfacial forces. These forces acting on fluids can be divided into normal and tangential components. In multiphase microchannel flows, viscous forces primarily act tangentially to the liquid–gas interface, elongating it, while interfacial forces mainly act normally to the interface, inducing droplet formation. The capillary number is used to express the competition between these two types of forces as [24]

$$Ca = \frac{\mu u}{\sigma}, \tag{5.1}$$

where μ is the viscosity, u is the mean velocity of the carrier fluid (continuous phase), σ is the interfacial tension.

With the chosen liquids, the capillary number depends solely on the mean velocity of the carrier fluid. For flows with low capillary numbers (typically $Ca < 10^{-2}$), the breakup of immiscible threads at the T-junction is primarily governed by interfacial forces rather than shear stresses. In this regime, shear stress has minimal effect on interface distortion as the dispersed phase enters the main microchannel. Consequently, the dispersed phase nearly fills the entire cross-section of the main channel. This forces the carrier fluid to flow only as thin wetting films along the microchannel walls, leading to pressure buildup upstream of the forming droplet. This pressure buildup leads to the narrowing of the immiscible thread's neck and eventually forms a plug at the T-junction.

Another significant characteristic of a fluid is its contact angle, which becomes apparent when a liquid droplet touches both a solid and a gas (or a solid and another immiscible liquid), as illustrated in Fig. 5.1. The liquid curvature results from differences in interfacial tensions. Young's equation for force balance is expressed as

$$\sigma_{sg} = \sigma_{sl} + \sigma_{lg} \cos \theta, \tag{5.2}$$

where σ indicates the interfacial tension between two phases, with subscripts g, l, s representing gas, liquid and solid, respectively, and θ as the contact angle. Therefore, in a microchannel, concave and convex interfaces are created upon filling, depending on whether the channel wall is hydrophobic or hydrophilic, as shown in Fig. 5.1. Hydrophobic channel surfaces are essential to prevent unstable, polydisperse water-in-oil droplet formation due to wetting effects.

The typical droplet formation process in a T-junction is as follows: Two immiscible fluids create an interface at the junction of the inlet and main channel. The flow of the dispersed phase penetrates into the main channel, initiating droplet growth. The pressure gradient and flow in the main channel distort the droplet downstream. The upstream interface of the droplet then moves downstream. As the interface approaches the downstream edge of the dispersed phase inlet, the neck connecting the inlet channel to the droplet breaks. The detached liquid plug flows downstream

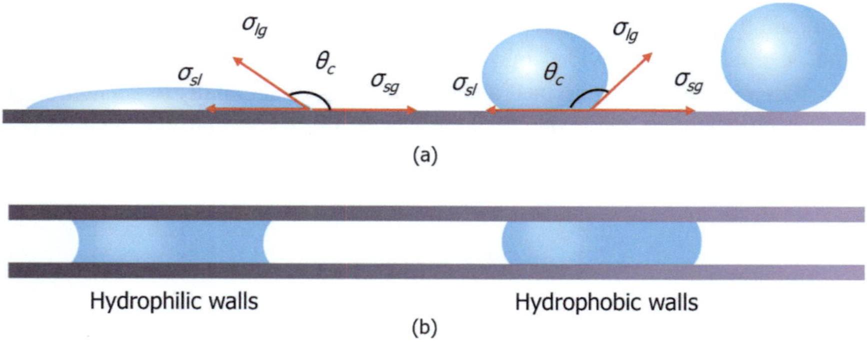

Fig. 5.1 Wettability of aqueous droplet: **a** on free surfaces and **b** within microchannels

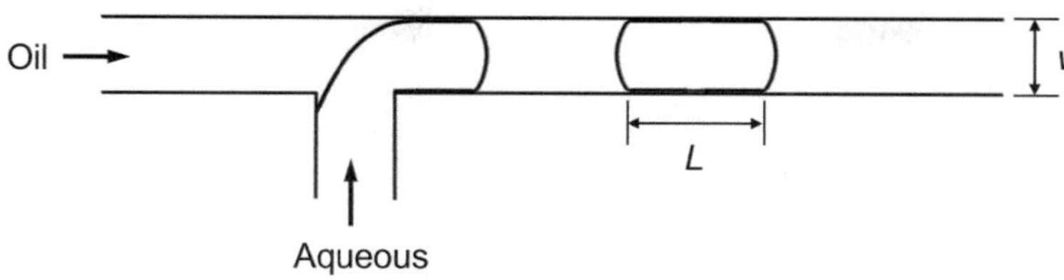

Fig. 5.2 Diagram of a T-junction microchannel used for droplet generation

in the main channel, while the tip of the dispersed phase stream retracts to the end of the inlet, repeating the process.

The geometry of the T-junction is illustrated in Fig. 5.2, where two channels merge at a right angle. The main channel carries the continuous fluid, while the orthogonal channel supplies the dispersed fluid. Both channels have rectangular cross-sections. The geometrical parameters defining the size and shape of the T-junction are the width of the main channel w and the height of the channel h, with L representing the length of the immiscible slug. Q_d and Q_c are the flow rates of the dispersed and continuous fluids, respectively.

The size of the droplet can be predicted quantitatively. Initially, the tip of the dispersed phase enters and blocks the main channel. At this point, the droplet's length is roughly equal to the channel's width. The increased pressure in the continuous fluid upstream of the immiscible thread begins to squeeze the neck (with a characteristic width of d). This causes the neck of the immiscible thread to narrow, eventually forming a plug at the T-junction. The size of the plug (L) is determined solely by the ratio of the flow rates of the two immiscible liquids and is expressed as

$$L = w\left(1 + a\frac{Q_d}{Q_c}\right), \tag{5.3}$$

where w is the width of the channel, Q_d and Q_c are the flow rates of the dispersed and continuous fluids respectively, and a is a constant dependent on the geometry of the T-junction. This scaling relation is independent of the fluid's material parameters, viscosities, and interfacial tension between them [25]. When the generated plug flows into a sudden expansion in the microfluidic channel, it transforms into a circular droplet. The diameters of the circular droplets can be calculated based on the volume of the plug.

In a typical setup, immersion oil is used as the continuous liquid and a $CaCl_2$ solution as the dispersed liquid for droplets. When a light source is placed above the chip, the droplet array effectively acts as a phase grating, with alternating regions of immersion oil and $CaCl_2$ droplets, each having different refractive indices, n_{oil} and $n_{droplet}$, respectively. The refractive index of the microdroplet can be precisely controlled. It is commonly noted the refractive index of a solution increases linearly with concentration. Adjusting the refractive index of the microdroplet is an efficient way to modify the optical properties of the optofluidic grating. For instance, to achieve a high optical performance in a liquid long-period grating, the chosen liquids must exhibit low light absorption or high transmission at the detection wavelength. The absorptions of the selected liquids were measured, and the results are presented in Table 5.1. The concentrations of $CaCl_2$ solutions are 1.60 mol/L ($n = 1.372$) and 4.16 mol/L ($n = 1.426$), respectively. The results indicate that nearly 90–99% of the light energy is absorbed in DI water and $CaCl_2$ solution at 1550 nm. However, for the other case, the absorption is reduced from 99 to 50% when a 1275 nm light source is used.

Figure 5.3 illustrates the droplet generation process. At the T-junction, the injection of $CaCl_2$ solution causes the interface between the two immiscible fluids to penetrate the main channel. Consequently, the $CaCl_2$ solution enters the main channel and gradually fills its entire cross-section. The immersion oil faces higher resistance, building up pressure until it breaks the neck of the elongated plug. The plug then moves downstream freely while the tip of the $CaCl_2$ solution retracts, ready for the next plug generation. When the generated plug reaches a sudden expansion in the microfluidic channel, it transforms into a circular droplet.

By coordinating the flow rates of two immiscible liquids, the droplets have diameters ranging from 20 to 50 μm, as shown in Fig. 5.4. The period can range from 10 to 50 μm simultaneously. The structure of the droplet grating can be dynamically and reversibly tuned by adjusting the flow rates. The diameter of a single droplet changes almost linearly within less than two seconds by altering the flow rates applied to

Table 5.1 Light absorption of different liquids ($\lambda = 1275$ nm and 1550 nm)

	DI water	$CaCl_2$ ($n = 1.372$)	$CaCl_2$ ($n = 1.426$)	Oil ($n = 1.462$)
1275 nm	0.52	0.49	0.48	0.05
1550 nm	0.99	0.93	0.91	0.34

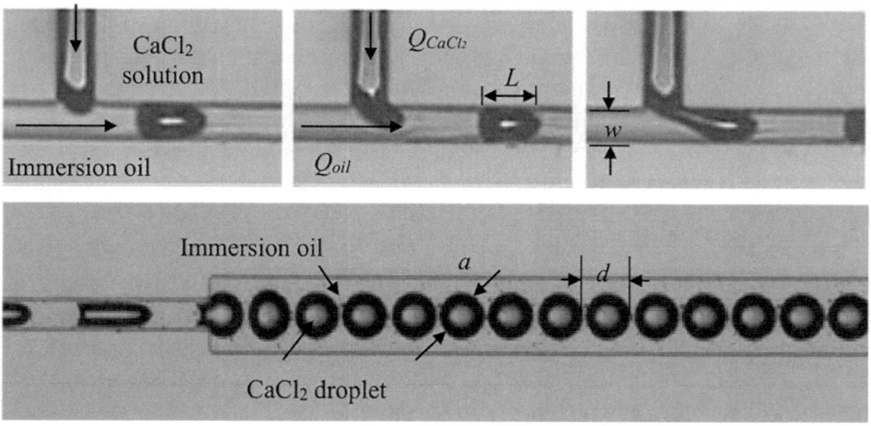

Fig. 5.3 Micrographs of the generation of droplets and the formation of droplet array. Reproduced from [14] with permission from Optical Society of America

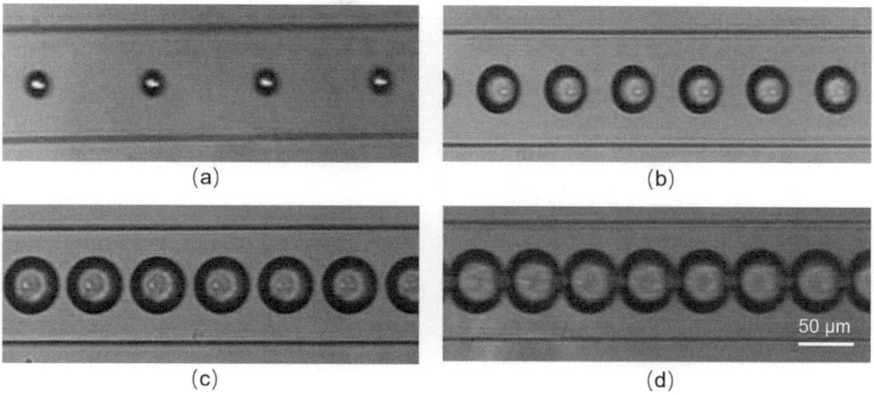

Fig. 5.4 Different droplet sizes and grating periods achieved by adjusting the flow rate

the device. As the flow rate of continuous phase varies, the grating period changes according to the droplet diameter. Additionally, the stability and monodispersity of the droplets are investigated. The standard deviations of the droplet size and grating period under certain flow rate ratios are measured and found to be less than 5%, indicating high uniformity and monodispersity. This makes it easy to form an array of droplets with constant size and period in the microchannel. The entire liquid grating system exhibits high stability over several hours when properly designed and operated.

5.3 Reconfigurable Optical Diffraction of Droplet Grating

Diffraction gratings

Diffraction gratings are typically categorized based on several criteria. The most intuitive distinction is between reflection and transmission gratings. They can also be classified as phase gratings and relief gratings. Phase gratings consist of refractive index variations, while relief gratings feature a surface relief structure that separates two media with different optical properties. Reflection gratings are usually relief gratings coated with highly reflective material. Reflectivity can vary significantly depending on the spectral region, making material choice critical for grating performance. Transmission gratings are often made with triangular, sinusoidal, or rectangular groove shapes. Triangular or blazed grooves are designed to direct as much light as possible of a specified wavelength band into one of the lower orders, typically the first, or to achieve a specific ratio of first to zero order at a given wavelength. At normal incidence, the symmetry of rectangular grooves leads to equal energy distribution in both plus and minus first orders, with the zero-order fraction at one wavelength varying from nearly 0 to over 90%. They are generally considered

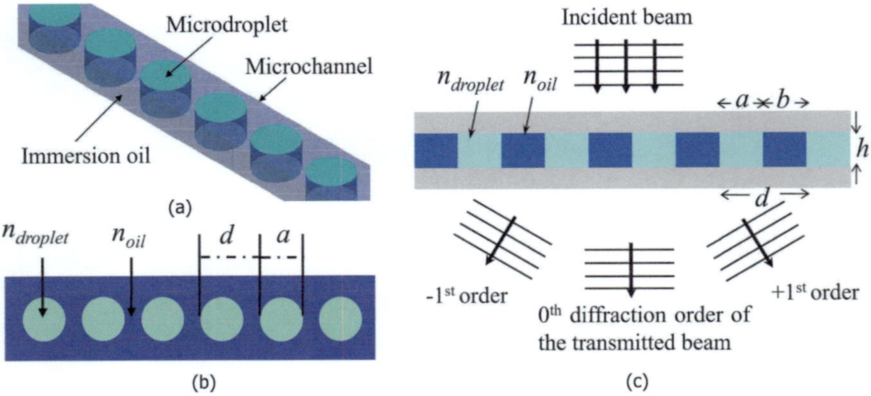

Fig. 5.5 **a** Schematic of the droplet optofluidic grating. **b** Top view of the droplet optofluidic grating. **c** Cross view of the droplet optofluidic grating. Reproduced from [14] with permission from Optical Society of America

phase gratings because their behavior is controlled by the physical phase retardation between light originating from successive grooves.

In an optofluidic grating, the grooves are replaced by multiphase droplets, as shown in Fig. 5.5. These droplets flow along the microchannel, forming a one-dimensional droplet array. When a light source is placed above the chip, this droplet array effectively functions as a phase grating, consisting of alternating regions of immersion oil and microdroplets, each with distinct refractive indices.

With constant refractive indices of the two immiscible liquids, the diffraction pattern of the droplet array combines single circular aperture Fraunhofer diffraction and multiple-slit interference, as shown in Fig. 5.6. The circular diffraction pattern is defined by the diameter of a single droplet. The irradiance distribution (I) follows the diffraction equation for a single circular aperture [26], which can be written as a function of θ:

$$I(\theta) = I_0\left(\frac{2J_1(ka\sin\theta)}{ka\sin\theta}\right)^2, \tag{5.4}$$

where a represents the diameter of a single droplet, k is the propagation number of incident light, J_1 is the first-order Bessel function. I_0 is the irradiance when $\theta = 0$, corresponding to the principal maximum. The period (i.e., radius of the 1st, 2nd and 3rd maxima) is inversely proportional to the diameter of the droplet.

The vertical pattern is determined by the droplet array dispersed in immersion oil, which creates periodic changes in refractive index, i.e., the droplet array functions as a phase grating. For any type of grating, the foundational grating equation applies

$$d\sin\theta_m = m\lambda, \tag{5.5}$$

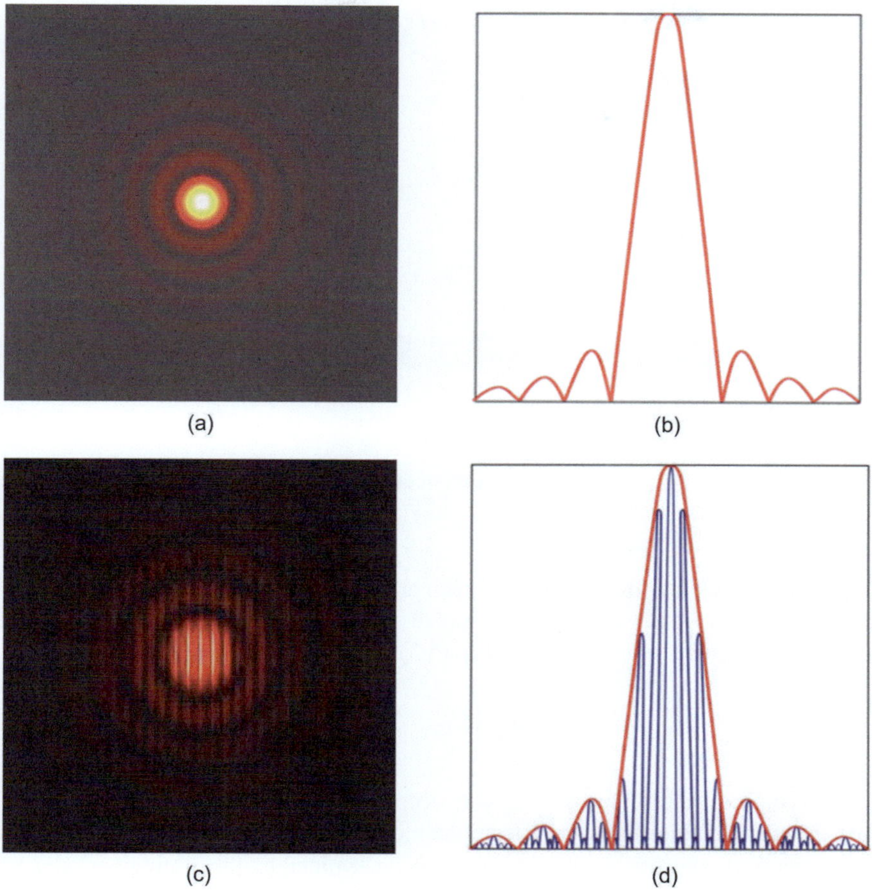

Fig. 5.6 Diffraction pattern analysis of the droplet grating: **a** circular diffraction pattern and **b** intensity distribution of single droplet; **c** vertical diffraction pattern and **d** intensity distribution of droplet array

where d is the grating period, λ is the light wavelength, and θ_m is the diffraction angle of the m-th order. The period of the vertical diffraction pattern (i.e., the spacing between two consecutive maxima) is inversely proportional to the grating period length:

$$\Delta y \approx s(\theta_m - \theta_{m-1}) = \frac{s\lambda}{d} \propto \frac{1}{d}, \tag{5.6}$$

where s is the distance between the grating and screen where the diffraction pattern is observed.

If the incident beam contains multiple wavelengths instead of a single wavelength, the droplet-based transmission grating also exhibits a dispersion effect. For a fixed

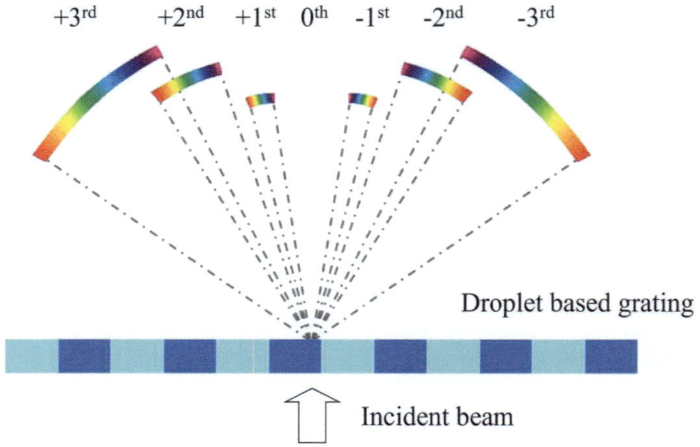

Fig. 5.7 Angular spread of the first three orders of the spectrum for the droplet-based optofluidic diffraction grating

direction of incidence, the direction of each principal maximum changes with wavelength. Figure 5.7 shows the angular spread of the continuous spectrum of visible light for the grating, as described by Eq. (5.5). From this, it results that

$$\theta_m = \sin^{-1}\frac{m\lambda}{d}. \tag{5.7}$$

Overlapping of different orders can occur when the product equals multiple possible combinations of m and λ from the incident light. The extent of this overlap also depends on the grating density, which is determined by the grating period d.

Phase transmission optical properties

The droplet-based phase transmission grating comprises alternating regions of immersion oil and droplets, each with different refractive indices, n_{oil} and n_{droplet}. The periodic phase shift ξ is given by

$$\xi = \frac{2\pi h(n_{\mathrm{oil}} - n_{\mathrm{droplet}})}{\lambda}, \tag{5.8}$$

where h is the height of the microfluidic channel (i.e., the thickness of the grating structure), λ is the wavelength of the incident beam, and $(n_{\mathrm{oil}} - n_{\mathrm{droplet}})$ denotes the refractive index contrast of the two liquids. The droplet profile also influences the phase shift. When the contact angle is 90°, the droplet exhibits an ideal rectangular cross-section profile, as shown in Fig. 5.7. As the contact angle decreases, the droplet profile becomes more rounded, distorting the wavefront phase shift profile. Such distortion can affect the diffraction patterns. For small droplets, the profile may be nearly circular. This discussion focuses on the ideal condition, although the design applies to other profiles as well.

Suppose that the total length of the grating is L, the transmission function of a phase grating $t(x)$ can be considered as the sum of two functions, $t_1(x)$ and $t_2(x)$, which are expressed as

$$t(x) = [t_1(x) + t_2(x)]\text{rect}\left(\frac{x}{L}\right), \tag{5.9}$$

where

$$t_1(x) = \frac{1}{d}\text{rect}\left(\frac{x}{a}\right)\text{comb}\left(\frac{x}{d}\right)\exp\left(\frac{j2\pi h n_{\text{droplet}}}{\lambda}\right), \tag{5.10a}$$

and

$$t_2(x) = \frac{1}{d}\text{rect}\left(\frac{x - d/2}{b}\right)\text{comb}\left(\frac{x}{d}\right)\exp\left(\frac{j2\pi h n_{\text{oil}}}{\lambda}\right). \tag{5.10b}$$

It is well-known that the field distribution of Fraunhofer diffraction of an aperture is the Fourier transform of the amplitude distribution across the aperture. When a plane wave normally incidents on the grating, the amplitude distribution across the grating can be described by the transmission function $t(x)$ of the grating. For convenience, assume $a = b = d/2$. Applying the Fourier transformation to Eq. (5.9), we have

$$T(f_x) = 2a\left\{\sum_{n=-\infty}^{+\infty} \begin{array}{l} \text{sinc}\left(\frac{m}{2}\right)\delta(df_x - m) \times \exp\left[j\pi\left(\frac{h\left(n_{\text{droplet}} + n_{\text{oil}}\right)}{\lambda} - \frac{m}{2}\right)\right] \\ \times \cos\left[\frac{\xi}{2} + \frac{m\pi}{2}\right] \end{array}\right\}$$
$$L\text{sinc}(Lf_x) \tag{5.11}$$

Thus, the intensity distribution of the Fraunhofer diffraction pattern of the phase grating can be represented as

$$I(f_x) = L^2 \sum_{n=-\infty}^{+\infty} \text{sinc}^2\left(\frac{m}{2}\right)\cos^2\left(\frac{\xi}{2} + \frac{m\pi}{2}\right) \times \text{sinc}^2\left[L\left(f_x - \frac{m}{d}\right)\right]. \tag{5.12}$$

When $m = 0$, Eq. (5.12) becomes

$$I_0(f_x) = L^2\cos^2\left(\frac{\xi}{2}\right) \times \text{sinc}^2(Lf_x). \tag{5.13}$$

This equation represents the intensity distribution of the zero-diffraction order. Consequently, the peak intensity of the zero-diffraction order is

$$I_{0\text{peak}} = L^2 \cos^2\left(\frac{\xi}{2}\right). \tag{5.14}$$

Similarly, if $m = 1$, we have

$$I_1(f_x) = L^2 \frac{4}{\pi^2} \sin^2\left(\frac{\xi}{2}\right) \times \text{sinc}^2\left[L\left(f_x - \frac{1}{d}\right)\right], \tag{5.15}$$

$$I_{1\text{peak}} = L^2 \frac{4}{\pi^4} \sin^2\left(\frac{\xi}{2}\right). \tag{5.16}$$

Since L is a constant, Eqs. (5.15) and (5.16) can be simplified as

$$I_{0\text{peak}} = \cos^2\left(\frac{\xi}{2}\right), \tag{5.17a}$$

and

$$I_{1\text{peak}} = \frac{4}{\pi^4} \sin^2\left(\frac{\xi}{2}\right). \tag{5.17b}$$

In the same way, the peak intensities of the higher diffraction orders can be derived by setting $m = 2, 3, \ldots, n$. It is observed that most diffraction energy is concentrated in the zero and first orders. The peak intensities of both the zero and first orders vary with the grating thickness h, the refractive indices difference between two immiscible liquids ($n_{\text{oil}} - n_{\text{droplet}}$), and the incident wavelength λ. For monochromatic incident light, changes in the phase shift ξ cause the diffraction efficiency of the zero and first orders to oscillate periodically, as shown in Fig. 5.8a. For broadband visible incident light, different wavelengths are transmitted at varying efficiencies, and the diffracted wavelength is filtered out from the 0th order. Since all wavelength components of the 0th order overlap at the same position, filtering out one or more wavelength components results in a subtractive color modification of the output spectrum, as shown in Fig. 5.8b.

Figure 5.9 illustrates the diffraction patterns generated by the droplet grating using a 632.8 nm He–Ne laser as the monochromatic incident source. The upper panels display the simulation results, while the lower panels show the experimental results. These images present the irradiance distribution for a droplet array grating diffraction pattern with varying droplet diameters and grating periods. The period of the circular pattern decreases as the droplet diameter increases. Conversely, the period of the vertical pattern increases as the grating period decreases. When the droplet size is close to the incident light wavelength, the diffraction pattern becomes distinctly visible.

Figure 5.10 illustrates the relationship between the diffraction pattern periods and the characteristics of the droplet grating. The first-order period of the circular pattern ranges from 5 to 20 as the droplet diameter varies between 10 and 20 μm. The vertical

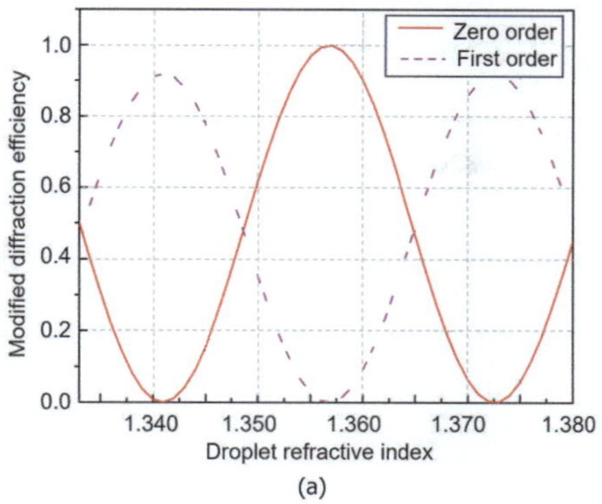

Fig. 5.8 **a** Simulation results of the diffracted efficiency of the zero-order and first-order diffraction as a function of droplet refractive index. **b** Simulation results of the spectrum shift of the zero-order with four different droplet refractive indices

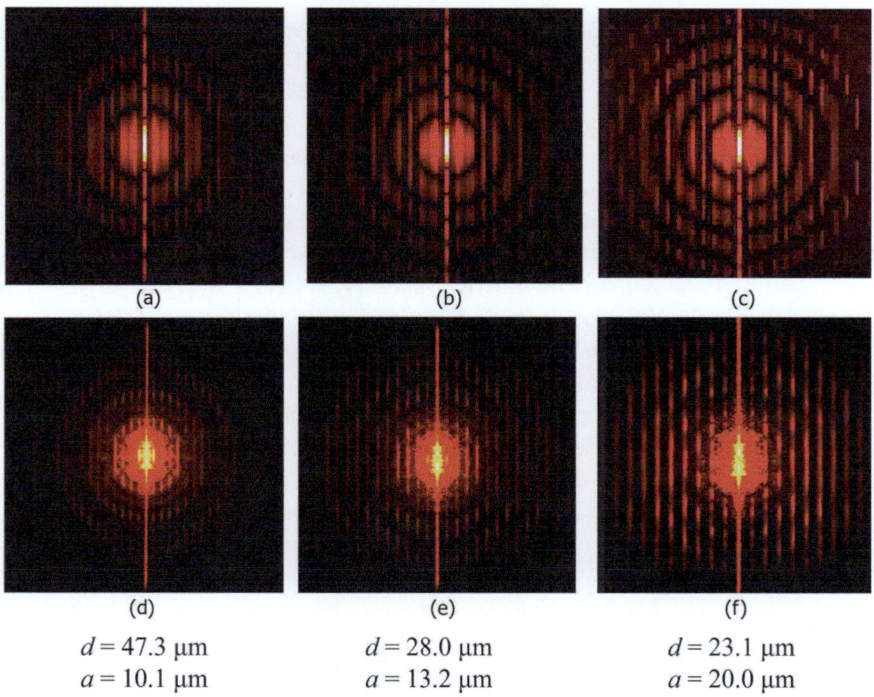

Fig. 5.9 Micrographs of the diffraction pattern of one-dimension droplet array with varying grating periods using monochromatic incident light: **a–c** Simulation results and **d–f** Experimental results. Reproduced from [14] with permission from Optical Society of America

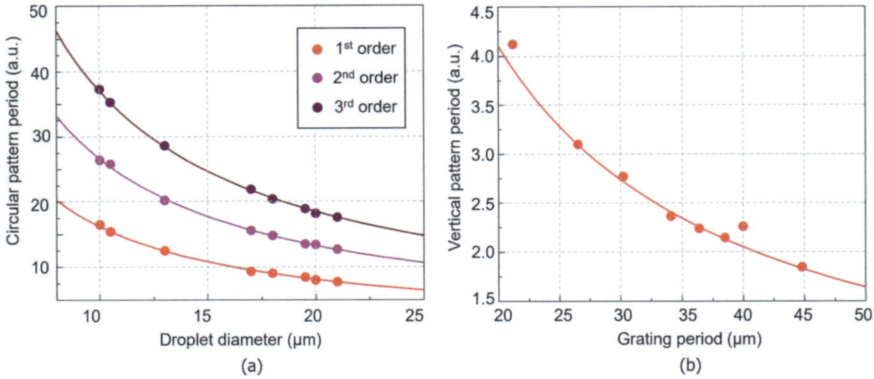

Fig. 5.10 **a** Relationship between the single droplet diameter and circular pattern period. **b** Relationship between the grating period and vertical pattern period

pattern period ranges from 1.5 to 4 as the grating period varies between 20 and 50 μm. The experimental results align well with the theoretical calculation curves.

The diffraction patterns vary with the refractive indices of the droplets for both monochromatic incident light and broadband white light. When the incident light is a single-wavelength laser (e.g., He–Ne laser), the intensities of the zero and first orders change with the values of n_{droplet}, as shown in Fig. 5.11a. The transmission efficiency varies almost linearly over the refractive index range from 1.344 to 1.352 and aligns well with theoretical predictions, as indicated in Fig. 5.8a. A total change in n_{droplet} of 0.016 refractive index unit (RIU), from 1.341 to 1.357, is measured within this intensity level range. The system's detection limit, or the change of RIU per unit change of intensity level, is 6.27×10^{-5} RIU. The channel height and the CCD camera resolution are the main factors influencing the detection limit. Therefore, the entire system can function as an optofluidic refractometer, which can be easily integrated into microdroplet devices like droplet-based microreactors and micromixers.

For white incident light, the color patterns of the 0th diffraction order are shown in Fig. 5.11b. Increasing n_{droplet} from 1.439 to 1.447 shift from magenta to red, clearly demonstrating the color filtering effect. As the refractive index of the solution changes, the central wavelength of the filtered color shifts at a rate of approximately 50 nm/RIU. This shows that the color filtering property can be finely tuned by adjusting the refractive index of the droplets, offering high sensitivity and stability.

5.4 Tunable Transmission of Optofluidic Long-Period Grating

Long-period grating (LPG) has shown promising applications in optical sensors. An LPG consists of a periodic perturbation of the refractive index, typically with a period of several hundred micrometers and a length ranging from several millimeters to

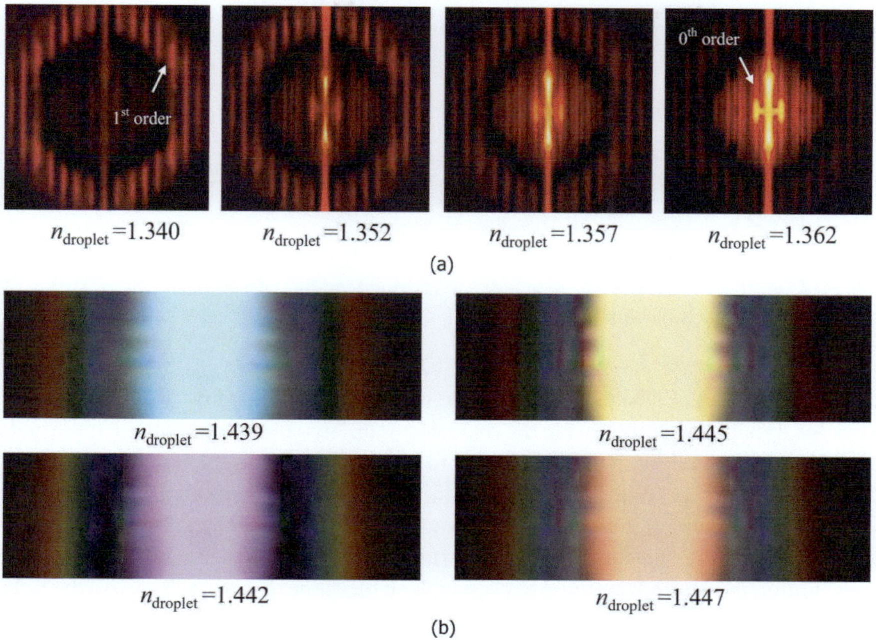

Fig. 5.11 **a** Variations of the diffraction patterns in response to the refractive index of droplets. **b** Measured diffractions of white incident light when the droplet grating is used as a 0th order color filter. Reproduced from [14] with permission from Optical Society of America

centimeters. It couples propagating light from core mode to cladding modes, leading to a series of attenuation bands centered at discrete wavelengths in the transmission spectrum. The central wavelength and attenuation strength of these bands depend on the index variation, period, and length of the gratings. For sensor applications, fiber LPGs are used as temperature, strain, and refractive index sensors [27]. Despite their broad applications, the performance of fiber LPGs is significantly limited by the constraints of optical fiber geometry and material. To overcome these limitations, LPGs have been realized in planar optical waveguide structures [28]. Optical waveguides can be fabricated using various materials and geometries, providing diverse design and application possibilities. However, LPGs in solid waveguides have limitations in tunability. For example, tuning the central wavelengths of the attenuation bands typically requires temperature adjustments, which may not be suitable for applications with strict temperature requirements, such as biosensors. Thus, incorporating a liquid waveguide in droplet microfluidics enables the realization of a widely tunable LPG.

An example of optofluidic long-period grating is shown in Fig. 5.12. In this optofluidic grating design, the two-phase plugs within the microchannel serve as the core layer, while the chip material functions as the cladding layer. With the refractive index of the oil phase and the aqueous phase to be higher than that of the

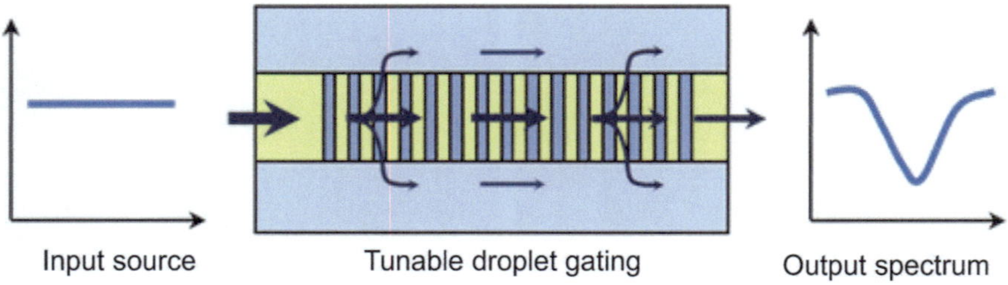

Fig. 5.12 Working principle of optofluidic long-period grating. Reproduced from [13] with permission from the Royal Society of Chemistry

chip material (e.g., PDMS), a microfluidic waveguide is created by droplet microfluidics for light propagation inside the microchannel. Two fibers are aligned at the ends of the microchannel, with one connected to a light source and the other to an optical detector or analyzer. When a broadband input is used, the output spectrum displays an attenuation band at a discrete wavelength. As light propagates through the core and encounters the optofluidic grating, certain distinct wavelengths are coupled with the cladding modes and continue to propagate in the same direction. Because the light in the cladding layer cannot be coupled back into the core due to high attenuation rates in the cladding modes, resonant attenuation bands appear in the transmission spectrum, as shown in Fig. 5.12. The central wavelengths of these attenuation bands (λ_i) are derived using the phase-matching condition of coupled-mode theory:

$$\lambda_i = \left(n_{\text{core}} - n^i_{\text{clad}}\right)d, \quad i = 1, 2, 3, \ldots \tag{5.18}$$

where d is the grating period, and n_{core} and $n^i{}_{\text{clad}}$ are the effective refractive indices of the core mode and the i_{th} cladding mode, respectively. In the optofluidic grating, the central wavelength can be easily tuned by adjusting the period of the droplet grating using different ratios of the flow rates of the two immiscible liquids. Additionally, the attenuation strength depends on the refractive index variation in the core (the difference between the refractive indices of the oil and aqueous phases) and the grating length.

The effect of shape profile of the plugs

The efficiency of light confinement in the microchannel is assessed against different shape profiles of the formed plugs. Four distinct shape profiles are simulated, as illustrated in Fig. 5.12. In the simulation, the microchannel length is 7 mm, the period is 100 μm, and the refractive indices of the continuous and dispersed phases are 1.462 and 1.430, respectively. The ideal shape profile is a plug with a 90° contact angle, identical to a solid grating, as shown in Fig. 5.13a. In this ideal condition, light is perfectly confined within the microchannel, and the grating effect is maintained. When the contact angle is reduced to 120°, light confinement within the microchannel remains effective, as depicted in Fig. 5.13b. However, light confinement deteriorates when the plug has a contact angle of 135°. In this case, light diverges into the cladding

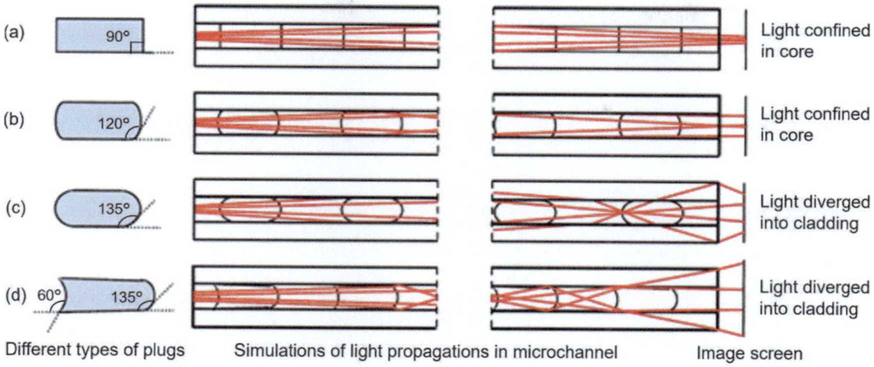

Fig. 5.13 Light propagation in microchannel with different types of plugs. **a** Flat curvature, **b** contact angle of 60°, **c** contact angle of 45° and **d** rear concave curvature. Reproduced from [13] with permission from the Royal Society of Chemistry

layer after propagating through five plugs, as shown in Fig. 5.13c. This leads to optical loss, and the light cannot be detected after it is reflected back. Figure 5.13d presents another shape profile with a concave curvature at the rear interface. This shape profile also fails to confine light in the core, similar to the one in Fig. 5.13c. Therefore, to ensure effective light confinement in the microchannel, the plugs should have a contact angle of approximately 90°.

The contact angle of the $CaCl_2$ plug in the PDMS microchannel filled with immersion oil is determined by the surface tensions at the interfaces between the three phases. When the water plug forms a boundary with the oil, its geometry adjusts to balance all the forces acting at the interface, eventually stabilizing with a specific contact angle θ. The forces acting on the interface are illustrated in Fig. 5.14. F_{sw}, F_{wo} and F_{os} represent the surface tension forces between the solid-water interface, water–oil interface, and oil-solid interface, respectively. F_a is the adhesive force between the liquid and solid surface. By balancing the forces in the horizontal direction, a relationship between the contact angle and the surface tension forces can be defined as

$$F_{sw} - F_{os} = -F_{wo}\cos\theta. \tag{5.19}$$

Given that the surface tension forces are directly proportional to their respective surface tensions, the relationship can be expressed as

$$\sigma_{sw} - \sigma_{os} = -\sigma_{wo}\cos\theta. \tag{5.20}$$

To form a convex interface with a contact angle greater than 90°,

$$\sigma_{wo} > \sigma_{sw} - \sigma_{os} > 0. \tag{5.21}$$

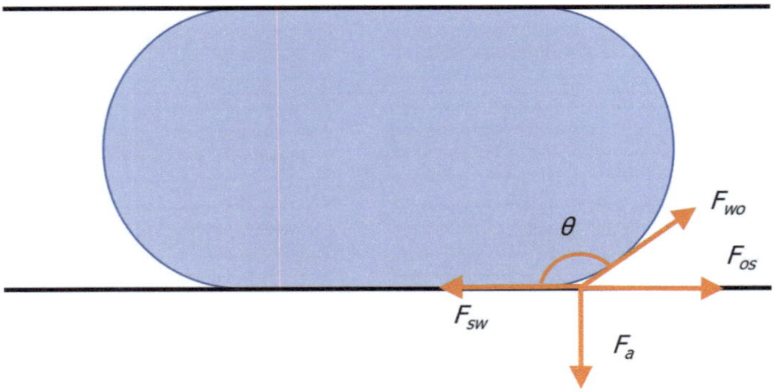

Fig. 5.14 Diagram of the forces acting on the boundary of the three phases

Conversely, to form a concave interface with a contact angle less than 90°,

$$\sigma_{wo} > 0 > \sigma_{sw} - \sigma_{os}. \tag{5.22}$$

To achieve an ideal grating with flat curvature (i.e., $\theta = 90°$), the solid–liquid surface tension must match the oil-solid surface tension ($\sigma_{sw} = \sigma_{os}$). This means the PDMS surface should be neither too hydrophilic nor too hydrophobic, providing partial wetting against the $CaCl_2$ phase. Several factors can affect the hydrophobicity of the PDMS wall. First, the surface roughness of the PDMS wall usually increases hydrophobicity [29]. Second, the PDMS wall becomes more hydrophilic after oxygen plasma treatment for chip bonding [30]. The hydrophobicity of PDMS gradually restores when the bonded chip is left in open air for several days. Additionally, the contact angle can be lowered by adding surfactants to reduce the surface tension of the aqueous solution.

Figure 5.15 illustrates the formation of an optofluidic grating within the microchannel. The immersion oil and $CaCl_2$ solution, injected into the microchannel, act as the continuous and dispersed phases, respectively. When the $CaCl_2$ solution disperses into the continuous phase, it significantly obstructs the flow of immersion oil, causing pressure to build up at the junction. A $CaCl_2$ plug forms when the pressure is sufficiently high. This process repeats, creating a long stream of plugs with a constant period and length. According to high-resolution images, the plugs' contact angle is approximately 115°–118°, as shown in Fig. 5.15. This ensures effective light confinement within the microchannel, as previously discussed in this section.

Tunable optical characteristics of optofluidic long-period grating

The length and period of the plugs can be adjusted by varying the continuous flow rate, as shown in Fig. 5.16a. The average measurement values are based on 30 measurements. By maintaining the $CaCl_2$ solution flow rate at 0.05 μL/s, increasing the oil flow rate from 0.035 to 0.05 μL/s reduces the $CaCl_2$ plug length from 47 to 36 μm, while the grating period increases from 72 to 87 μm at a constant rate. Conversely, the length and period of the plugs can also be modified by varying the

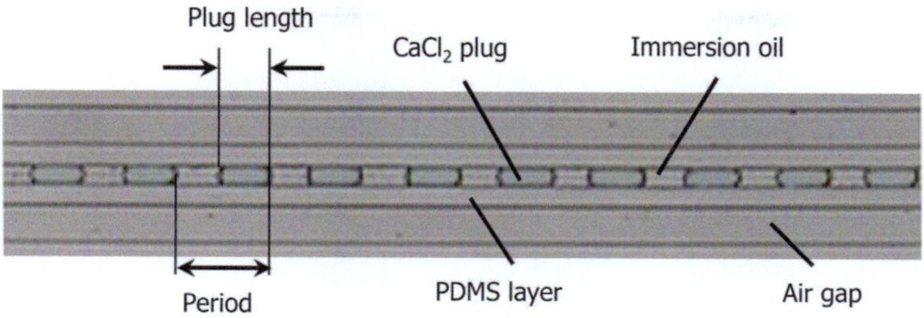

Fig. 5.15 Micrographs of optofluidic long-period grating in the microchannel. Reproduced from [13] with permission from the Royal Society of Chemistry

dispersed flow rate, as shown in Fig. 5.16b. By keeping the oil flow rate at 0.05 μL/s, increasing the $CaCl_2$ solution flow rate from 0.05 to 0.065 μL/s results in an increase in both the $CaCl_2$ plug length and the grating period. The standard deviations of the plug length and period are within 5%. As the plugs are highly uniform and monodisperse, these variations do not significantly affect the optical performance of the liquid grating.

The optical properties of the optofluidic grating are influenced by the grating period and length. Figure 5.17a shows the normalized transmission spectra for a grating length of 6 mm with different periods of 70, 75, and 80 μm. The liquid grating is formed using immersion oil and $CaCl_2$ solution with refractive indices of 1.462 and 1.430, respectively, resulting in an index variation of 0.032. As the grating period increases, the attenuation peak shifts from 1264.2 nm to longer wavelengths of 1274.6 and 1281.5 nm. Additionally, the attenuation strength increases from 0.23 to 0.37 and 0.52. This is because the number of periods in the optofluidic grating with a longer period is relatively smaller compared to those with shorter periods.

Figure 5.17b illustrates the transmission spectra of the optofluidic grating with different grating lengths of 5, 6, and 7 mm. The grating period and index variation remain constant at 75 μm and 0.032, respectively. As a result, the attenuation peak

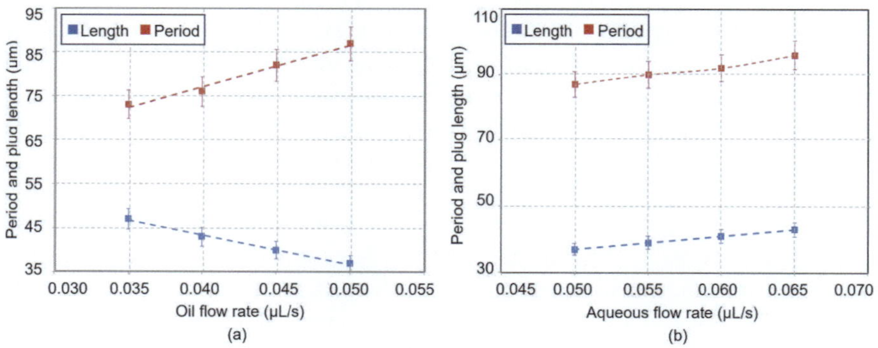

Fig. 5.16 **a** Tuning of grating period and plug length relative to the flow rate of the immersion oil. **b** Tuning of grating period and plug length relative to the flow rate of the aqueous $CaCl_2$ solution

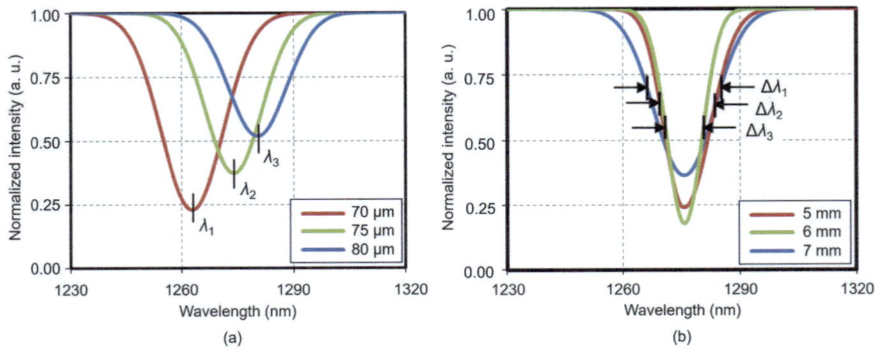

Fig. 5.17 Transmission spectra of the optofluidic grating with varying **a** grating periods and **b** grating lengths. Reproduced from [13] with permission from the Royal Society of Chemistry

stays constant at 1274.6 nm, but the attenuation strength is significantly affected. Increasing the grating length enhances the attenuation strength from 0.37 to 0.24 and 0.14. Consequently, the bandwidth of the attenuation band decreases from 10.2 to 9.0 and 7.6 nm. This indicates that a longer grating length improves the attenuation strength, leading to a sharper and narrower attenuation band.

The optofluidic long-period grating has demonstrated tunability in the central attenuation wavelength, attenuation strength, and attenuation bandwidth. The above example shows an attenuation strength of 0.14 and a bandwidth of 7.6 nm for a grating length of 7 mm and an index variation of 0.032. The sharpness of the attenuation band can be further improved with a longer grating period, though it is limited by detectable intensity due to liquid absorption. The interfacial curvature causes the oil–water interface to be less than ideal as a grating interface, slightly affecting the optical performance in terms of resonant wavelength and attenuation strength. These effects can be compensated for by adjusting the plug period and refractive index variation. Additionally, reducing interfacial curvature can be achieved through PDMS surface treatment and surfactant use.

The advantages of optofluidic grating realized by droplet microfluidics are remarkable. First, various liquid combinations can overcome material restrictions commonly seen in solid gratings. For example, in fiber grating interferometers, the modulation in the fiber core is limited to 0.0005 [31]. This can be enhanced to 0.001 using doping techniques [31] or hydrogen loading [32]. Unlike solid gratings, optofluidic gratings can easily achieve a high index modulation of 0.02 by selecting different liquids. Second, optofluidic gratings offer flexible options for input light sources and detectors. The grating periods and subsequent attenuation bandwidths can be easily tuned, providing a broad tunable range in the central resonant wavelength and attenuation strength, which is challenging for solid gratings. Finally, optofluidic gratings in microchannels can be easily integrated onto microfluidic chips, simplifying fabrication processes compared to solid gratings.

5.5 Summary

This chapter focuses on optofluidic gratings, which are reconfigurable using multiphase droplets on a microfluidic chip. These optofluidic droplet gratings boast high stability and wide tunability in droplet size, grating period, and refractive index, making them ideal for biochemical and biomaterial applications. To manage the more complex multiphase fluids compared to single-phase flow, various physical phenomena in microfluidic devices need to be analyzed. The formation process of droplets in the microfluidic T-junction is discussed here. At low capillary numbers, the dominant factor in the dynamics break-up is the pressure drop across the emerging droplet. A simple scaling relation predicts the size of droplets produced in T-junctions, considering the flow rates of the two immiscible phases, the viscosity of the continuous phase, the interfacial tension, and the device's geometrical dimensions. For optofluidic diffraction gratings, uniformly spaced circular droplets offer high tunability in grating period and refractive index variation, resulting in various diffraction patterns. A compact optofluidic refractometer can realized using the droplet grating system. The droplet grating also functions as a zeroth-order transmission color filter. For an optofluidic long-period grating, droplets with a 90° contact angle are preferred for effective light confinement. The design guidelines are based on liquid selection, droplet size, and shape profile. The optofluidic long-period grating demonstrates flexible attenuation peaks and strengths. By tuning the grating period and length, the optofluidic long-period grating can be configured to ensure the attenuation region appears in different wavelength ranges, making it suitable for a broad range of applications as a biosensor for chemical or biomedical analysis.

References

1. Teh SY, Lin R, Hung LH, Lee AP (2008) Droplet microfluidics. Lab Chip 8(2):198–220
2. Song H, Tice JD, Ismagilov RF (2003) A microfluidic system for controlling reaction networks in time. Angew Chem Int Ed 42(7):768–772
3. Chan EM, Alivisatos AP, Mathies RA (2005) High-temperature microfluidic synthesis of CdSe nanocrystals in nanoliter droplets. J Am Chem Soc 127(40):13854–13861
4. Chen DL, Gerdts CJ, Ismagilov RF (2005) Using microfluidics to observe the effect of mixing on nucleation of protein crystals. J Am Chem Soc 127(27):9672–9673
5. Zheng B, Ismagilov RF (2005) A microfluidic approach for screening submicroliter volumes against multiple reagents by using preformed arrays of nanoliter plugs in a three-phase liquid/liquid/gas flow. Angew Chem Int Ed 44(17):2520–2523
6. Chatterjee D, Hetayothin B, Wheeler AR, King DJ, Garrell RL (2006) Droplet-based microfluidics with nonaqueous solvents and solutions. Lab Chip 6(2):199–206
7. Wheeler AR, Moon H, Kim CJ, Loo JA, Garrell RL (2004) Electrowetting-based microfluidics for analysis of peptides and proteins by matrix-assisted laser desorption/ionization mass spectrometry. Anal Chem 76(16):4833–4838
8. Leamon JH, Link DR, Egholm M, Rothberg JM (2006) Overview: methods and applications for droplet compartmentalization of biology. Nat Methods 3(7):541–543
9. Liau A, Karnik R, Majumdar A, Cate JH (2005) Mixing crowded biological solutions in milliseconds. Anal Chem 77(23):7618–7625

10. Li ZY, Zhang Z, Emery T, Scherer A, Psaltis D (2006) Single mode optofluidic distributed feedback dye laser. Opt Express 14:696–701
11. Song W, Andreas EV, Li Z, Psaltis D (2009) Optofluidic evanescent dye laser based on a distributed feedback circular grating. Appl Phys Lett 94:161110
12. Chin LK, Liu AQ, Zhang JB, Lim CS, Soh YC (2008) An on-chip liquid tunable grating using multiphase droplet microfluidics. Appl Phys Lett 93:164107
13. Chin LK, Liu AQ, Soh YC, Lim CS, Lin CL (2010) A reconfigurable optofluidic Michelson interferometer using tunable droplet grating. Lab Chip 10:1072–1078
14. Yu JQ, Yang Y, Liu AQ, Chin LK, Zhang XM (2010) Microfluidic droplet grating for reconfigurable optical diffraction. Opt Lett 35(11):1890–1892
15. Wu L, Li GP, Xu W, Bachman M (2006) Droplet formation in microchannels under static conditions. Appl Phys Lett 89(14):144106
16. He MY, Kuo JS, Chiu DT (2005) Electro-generation of single femtoliter- and picoliter-volume aqueous droplets in microfluidic systems. Appl Phys Lett 87(3):031916
17. Kim SJ, Song YA, Skipper PL, Han J (2006) Electrohydrodynamic generation and delivery of monodisperse picoliter droplets using a poly(dimethylsiloxane) microchip. Anal Chem 78(23):8011–8019
18. Zeng J, Korsmeyer T (2004) Principles of droplet electrohydrodynamics for lab-on-a-chip. Lab Chip 4(4):265–277
19. Lorenz RM, Edgar JS, Jeffries GD, Chiu DT (2006) Microfluidic and optical systems for the on-demand generation and manipulation of single femtoliter-volume aqueous droplets. Anal Chem 78(18):6433–6439
20. Quevedo E, Steinbacher J, McQuade DT (2005) Interfacial polymerization within a simplified microfluidic device: capturing capsules. J Am Chem Soc 127(30):10498–10499
21. Christopher GF, Anna SL (2007) Microfluidic methods for generating continuous droplet streams. J Phys D Appl Phys 40(19):R319–R336
22. Baroud CN, Gallaire F, Dangla R (2010) Dynamics of microfluidic droplets. Lab Chip 10(16):2032–2045
23. Seemann R, Brinkmann M, Pfohl T, Herminghaus S (2012) Droplet based microfluidics. Rep Prog Phys 75(1):016601
24. Joanicot M, Ajdari A (2005) Droplet control for microfluidics. Science 309(5736):887–888
25. Garstecki P (2006) Formation of droplets and bubbles in a microfluidic T-junction-scaling and mechanism of break-up. Lab Chip 6(3):437–446
26. Hecht E (2017) Optics, 5th edn. Pearson Education
27. Bhatia V, Vengsarkar AM (1996) Optical fiber long-period grating sensors. Opt Lett 21:692–694
28. Liu Q, Chiang KS, Lor KP (2007) Dual resonance in a long-period waveguide grating. Appl Phys B Lasers Opt 86:147–150
29. Jin M, Feng X, Xi J, Zhai J, Cho K, Feng L, Jiang L (2005) Super-hydrophobic PDMS surface with ultra-low adhesive force. Macromol Rapid Commun 26:1805–1809
30. Bhattacharya S, Datta A, Berg JM, Gangopadhyay S (2005) Studies on surface wettability of poly(dimethyl) siloxane (PDMS) and glass under oxygen-plasma treatment and correlation with bond strength. J Microelectromech Syst 14:590–597
31. Douay M, Xie WX, Taunay T, Bernage P, Nilay P, Cordier P, Poumellec B, Dong L, Bayon JF, Poignant H, Delevaque E (1997) Densification involved in the UV-based photosensitivity of silica glasses and optical fibers. J Lightwave Technol 15:1329–1342
32. Poumellec B, Nilay P, Duoay M, Bayon JF (1996) The UV-induced refractive index grating in Ge:SiO_2 preforms: additional CW experiments and the macroscopic origin of the change of index. J Phys D Appl Phys 29:1842–1856

Chapter 6
Optofluidic Nanobubble and Nanodroplet

6.1 Introduction

In its early stages, bubble research was primarily driven by the problems they caused in marine and hydraulic systems. Researchers focused on minimizing the damage from violent bubble collapses and explored their dynamics within surrounding flows. Today, bubbles hold a significant place in modern science and technology due to their frequent presence in fluid systems and their applications across diverse fields such as biology, chemistry, physics, engineering, and medicine. The intriguing properties of bubbles, including nucleation, oscillation, and sonoluminescence, continue to captivate researchers.

The most well-known application of bubbles is ultrasonic cleaning. During bubble collapse, micro-streaming forms, effectively removing dirt particles from surfaces. In miniaturized devices, such as lab-on-a-chip and fuel cells, bubbles can be used as pumps to control microfluidics [1]. Recent advancements in medical applications involve bubbles in cell operations. They can pass through capillaries, releasing drugs and genes incorporated on their surface [2], and are also utilized for cell stretching, lysis, and membrane poration [3, 4].

Bubbles can be generated in two main ways, depending on the system. In an *n*-ary system, one or more components may be supersaturated and form bubbles, as in carbonated drinks. In a unary system, bubbles are generated by boiling or cavitation. Boiling involves rupturing a liquid by increasing its temperature at constant pressure. Cavitation involves rupturing a liquid by decreasing its pressure at constant temperature. In the context of optofluidics, cavitation bubbles can be generated by a focused laser pulse in fluid. This process, called optical breakdown, is a nonlinear absorption process that forms plasma at locations where the threshold irradiance is surpassed. Once high-pressure plasma is created, the surrounding liquid compresses and starts to flow outward, with the inertia of the external liquid confining it. A bubble forms and undergoes violent expansion. When the internal pressure drops below the external pressure, the bubble collapses.

S. Xiong and A.-Q. Liu, *Optofluidics: Fundamentals and Applications*,
Microfluidics and Nanophotonics: Science and Engineering 4,
https://doi.org/10.1007/978-981-95-6296-1_6

Lord Rayleigh was the first to mathematically describe bubble dynamics, laying the foundation for today's research on cavitation bubbles [5]. In the 1910s, he tackled the problem of an empty cavity collapsing in a large liquid mass, focusing solely on inertia. The following researchers explored bubble dynamics in more complex scenarios, ranging from vaporous to gaseous states, static to traveling, and spherical to arbitrary shapes. Practical considerations such as surface tension, liquid viscosity, thermal effects, liquid compressibility, gas compressibility, and mass diffusion were included. When a bubble is subjected to a periodic pressure field, its radius oscillates in response to the driving pressure, exhibiting specific nonlinear phenomena [6]. Additionally, understanding bubble–bubble and bubble-wall interactions is essential for many applications that require focused energy and flow in a small area. When a bubble is near a surface, its asymmetric collapse results in jetting flow, vortices, and a sequence of shock waves. The group behavior of bubbles is also intriguing, as they often appear in large numbers and arrange themselves in clusters, streamers, clouds, and swarms. Researchers have studied bubble distributions, bubble tracks, and velocity distributions [7, 8].

Despite extensive studies on bubble dynamics in macroscopic conditions, there is still limited knowledge about bubbles on an extremely small scale, such as nanobubbles. Jets can form from the asymmetric expansion and collapse of bubbles, which are crucial in bubble-related applications. As a result, nanojets have become an attractive topic, accompanied by complex nonlinear phenomena. This chapter explores the formation and dynamics of laser-induced micro/nanobubbles and their resulting jets, which also contribute to the creation of nanobubbles or nanodroplets.

6.2 Laser Induced Nanobubble

Study of Nanobubbles

Most theories on bubbles are based on macroscopic considerations, using standard continuum fluid mechanics. A significant limitation of continuum mechanical approaches is the assumption of local thermodynamic equilibrium, which presumes that macroscopic fluid variables do not change drastically over molecular length and time scales. However, it has been observed that classical hydrodynamic modeling fails to accurately represent bubbles in the final stages of collapse, particularly when they reach the nanometer scale.

As an alternative approach, molecular dynamics (MD) simulations are used to describe the process of bubble collapse [9]. For nanobubbles with radii of 1 and 1.5 nm, the collapse time is less than 10 ps and depends on temperature [10, 11]. A focused jet is observed during the shock-induced collapse of nanobubbles with diameters of 6–8 nm, investigated using MD simulations based on a reactive force field [12]. MD simulations can also be used to study the formation, stability, and breakup of nanojets. Recent advancements in molecular simulations and statistical mechanics of inhomogeneous fluids have sparked new interest in the study of bubble nucleation.

While nanobubble dynamics have been extensively studied through simulations, there are limited experimental methods to explore their properties. The existence of nanobubbles on hydrophobic surfaces was first reported in 1994 [13]. Significant progress in nanobubble research followed after images of surface nanobubbles were obtained using the Atomic Force Microscope (AFM). Today, surface nanobubbles are widely reported on various material surfaces using multiple detection mechanisms. Although AFM is the most commonly used technique, others such as rapid cryofixation/freeze fracture, optical scattering, and total internal reflection fluorescence (TIRF) are also employed. Typical surface nanobubbles are spherical caps with heights around 10 nm and diameters around 100 nm, located at the liquid–solid interface.

The most common method for producing surface nanobubbles is solvent exchange. A solvent that is miscible with water and has a higher gas solubility, such as ethanol, is used. The substrate is initially immersed in ethanol, which is later replaced by water. Due to ethanol's preferential wetting of hydrophobic surfaces, it resists complete removal. Upon mixing with water, gas is released because the solubility of gas in the aqueous phase is exceeded. Additionally, nanobubbles can be generated by heating the substrate. Surface nanobubbles differ from regular macroscopic bubbles; they are stable against violent decompression and have a remarkably long lifetime. According to classical diffusion theory, a gaseous bubble with a radius of 100 nm should disappear within 1 μs in degassed water. However, surface nanobubbles have been observed to last for more than 5 days [14]. Understanding their super stability has emerged as a key research focus. One possible explanation is that the diffusive outflux is stably balanced by an influx of gas at the contact line [15].

Femtosecond lasers are capable of creating nanosized cavitation bubbles in water. As the laser pulse duration decreases, the energy required for optical breakdown is also decreased. Consequently, both shock wave emission and cavitation bubble expansion are significantly diminished. Bubbles produced by longer laser pulses, such as nanoseconds and picoseconds, are typically tens or hundreds of micrometers in size. Ultrashort laser pulses are believed to form nanometer-sized cavitation bubbles by combining heat and thermoelastic tensile stress. Early studies show that bubbles produced by femtosecond laser pulses have maximum radii of 45–100 μm when the energy is 6–10 times above the threshold [16, 17]. Research has been conducted to determine the maximum bubble radius as a function of laser pulse energy. By detecting the scattering signals, the bubble oscillation time is measured, allowing for the deduction of the maximum bubble radius. At the threshold, the maximum bubble radius is smaller than the diffraction-limited focus radius, ranging from 190 to 320 nm [18].

Alternatively, strong local heating can create transient nanobubbles around nanoparticles. For example, gold nanoparticles with a diameter of 9 nm, when excited by a femtosecond laser, produce nanobubbles with a maximum radius of 20 nm and a lifetime within 1 ns [19]. Pulsed laser causes strong nonequilibrium heating of the particle and the surrounding water shell. The particle's surface temperature can easily exceed the boiling point of water. The power threshold for bubble formation is influenced by both the nanoparticle size and the aggregation state of the nanoparticles [20]. The presence of gold nanoparticles increases local absorption, significantly reducing the power and energy required to achieve similar bubbles without nanoparticles. Larger bubbles can be generated with higher power or longer pulse durations. The lifetime of a nanobubble depends on the nanoparticle size, aggregation state, and laser intensity, ranging from sub-nanoseconds to several hundred nanoseconds [21].

Femtosecond laser-induced nanobubbles, whether with or without nanoparticles, are small and unstable. They are difficult to detect directly using microscopes or cameras due to the diffraction limit. X-ray scattering is the most commonly used technique to investigate nanobubbles, providing a temporal depiction of the structural and thermal dynamics of nanobubbles and the surrounding medium. In these studies, bubble dynamics are still described using the Rayleigh-Plesset equation and continuum thermodynamics.

Nanobubbles were generated in water-filled hydrophilic nanochannels due to evaporation [22]. These nanochannels are etched on a silica substrate, measuring 120 μm in length and 4 μm in width, with depths ranging from 20 to 120 nm. The nucleation of the nanobubbles results from evaporation-induced negative pressure. Because these nanobubbles have a one-dimensional nanoscopic scale, their behaviors are directly observed through a microscope system. The vapor bubbles exhibit unusual motion, as well as translational stability and symmetry, due to the competition between evaporation-induced hydraulic flow and thermocapillary-driven flow.

Generating Nanobubbles by Laser

In the following section, we discuss the dynamics of nanobubbles generated by a focused laser spot. On the micrometer scale, dynamics may still be dominated by inertial forces, but as the size decreases, viscosity, surface tension, and thermo-capillary flow become more significant. Here, we aim to explore how dynamics are affected by sub-micrometer confinement. An experimental setup to achieve this confinement, which can be observed with high-speed imaging, is an extended nanochannel. This channel is etched in glass with a height of 300–900 nm and a width of several hundred micrometers.

The channels are fabricated in a borosilicate glass (e.g., Pyrex 7740) using standard photolithography and wet chemical etching processes. Due to its low coefficient of thermal expansion, borosilicate glass is ideal for high-temperature applications and suitable for thermal bonding. The chip fabrication processes are illustrated in Fig. 6.1.

The design layout of the optofluidic chip is created using a computer-aided design (CAD) program. A dark field clear feature chrome mask is fabricated on a piece of sodalime glass based on the drawing. For the combined microchannel and nanochannel chip, the masks for each are separate but share the same alignment marks. A thin layer of photoresist is spin-coated onto the glass plate, which is then soft-baked on a hot plate at 105 °C for 3 min. The glass is then exposed to UV light for 1–2 min under the mask, developed, rinsed with DI water, and dried with nitrogen gas. Then, the glass is hard baked at 105 °C for 60 min. Structures on

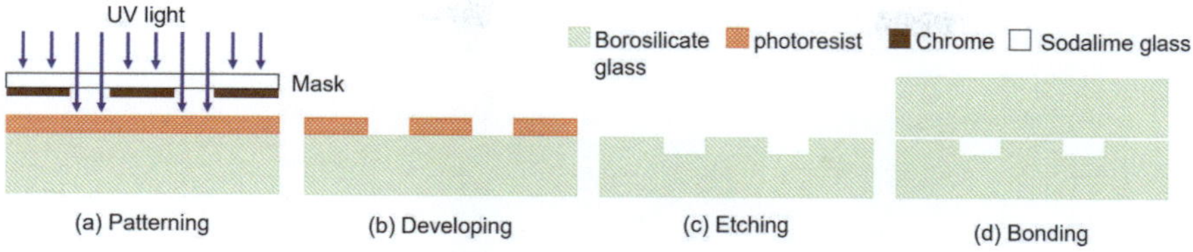

Fig. 6.1 Fabrication process flow of glass chip

the glass are etched using an HF/HCl solution. The etching depth is determined by the etching time and HF concentration. Adding HCl improves the process quality, resulting in a smoother etched surface. Different HF/HCl compositions and etching times are used for microchannels and nanochannels. After etching, the photoresist is removed with acetone, and the glass is rinsed with IPA and water. For chips with both nanochannels and microchannels, the microchannels are etched first, followed by a second photoresist coating before nanochannel fabrication. Inlets and outlets are drilled post-etching. The etched glass substrate and another cover glass are fused in a furnace at 625 °C for 2 h, resulting in a sealed micro/nanofluidic chip.

Figure 6.2 shows the experimental setup which consists of a pulsed Nd:YAG laser, a digital delay generator, an inverted microscope and a high-speed camera. The laser beam is expanded by the lenses L_1 and L_2, and reflected from mirrors into the microscope. The bubble is generated by focusing the laser pulse with the microscope objective. The energy of the laser pulse is controlled by delaying the interval between the flash lamp and the Q-switch trigger. The micro/nano-structured channel is filled with blue ink, which strongly absorbs the green light and yields a good contrast for the bright field illumination. High-speed recording is done at framing rates between 450,000–675,000 frame/s with the lowest possible exposure time of 370 ns.

Bubble Dynamics in Confined Channel

Conventional theories typically assume that bubbles maintain a spherical shape throughout the entire process, thereby enabling their dynamics to be described by the radius as a function of time. Rayleigh addressed the problem of a bubble collapsing in a liquid as [5]

$$R\ddot{R} + \frac{3}{2}\ddot{R}^2 = \frac{p}{\rho}, \tag{6.1}$$

where R, $\dot{R}$ and $\ddot{R}$ represent the bubble radius and its time derivatives, ρ is the liquid density, and p is the pressure in the liquid far from the bubble.

Plesset developed a more generalized equation, known as the Rayleigh-Plesset equation, which is expressed as [23]

$$R\ddot{R} + \frac{3}{2}\dot{R}^2 = \frac{1}{\rho}\left(p_B - p_\infty - \frac{2\sigma}{R} - \frac{4\mu}{R}\dot{R}\right), \tag{6.2}$$

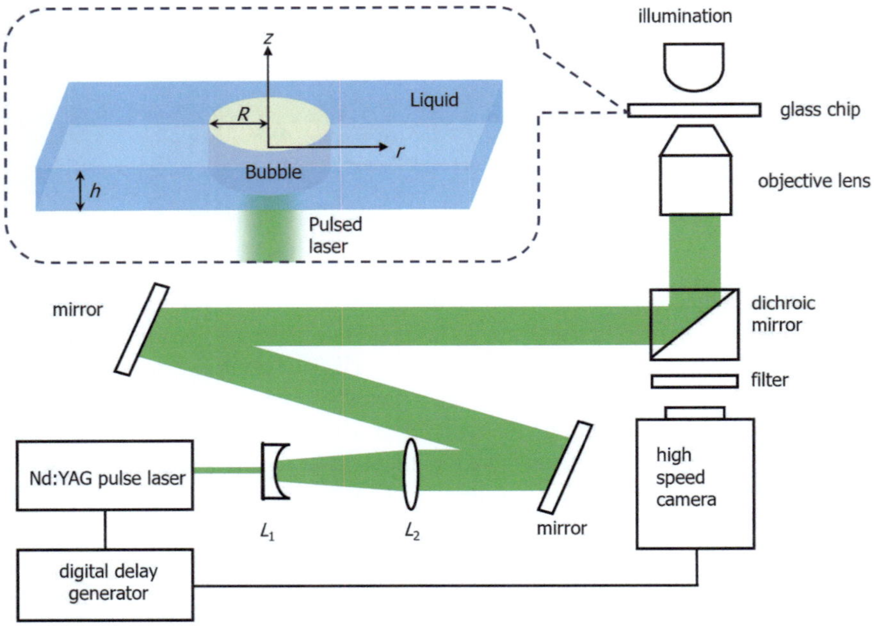

Fig. 6.2 Schematics of the experimental setup for laser-induced bubbles

where p_B is the pressure inside the bubble, p_∞ is the pressure infinitely far from the bubble, σ is the surface tension, and μ is the dynamic viscosity of the liquid. This equation, derived for Newtonian fluids from the Navier–Stokes equation, describes the entire evolution of the bubble from onset through expansion to collapse.

If the bubble forms in a shallow channel, where the height is much smaller than the bubble radius, it assumes a cylindrical shape due to the confinement by the top and bottom channel walls, as illustrated in the inset of Fig. 6.2. The planar bubble dynamics can be described as [24]

$$\left(R\ddot{R} + \dot{R}^2\right)\log\frac{R}{R_\infty} + \frac{1}{2}\dot{R}^2 = \frac{p}{\rho}, \tag{6.3}$$

where R is the bubble radius, ρ is the liquid density, p is the pressure in the liquid far from the bubble, and R_∞ is the distance at which the fluid velocity has dropped to zero. Here R_∞ is identified as half of the channel width, supposing the bubble is generated at the center of the channel.

Figure 6.3a illustrates the bubble evolution in an 8-μm high, 1-mm wide, and 3-mm long microchannel. Sudden absorption of the pulsed laser energy vaporizes the liquid. The hot vapor expands rapidly, reaching the maximum size of the microbubble in 2 μs. As the vapor temperature decreases, its pressure eventually drops below that of the surrounding liquid. The pressure imbalance between the inside and outside of the bubble causes it to collapse within 6 μs. The dynamics are undisturbed by other boundaries due to the significant distance between the bubble and the side walls. Consequently, the radial evolution of the microbubble is symmetrical. Since the microbubble radius (with a maximum value of 22.5 μm) is much larger than the channel height, it adopts a cylindrical shape confined by the top and bottom channel walls. Figure 6.3b demonstrates that the measured bubble radius closely matches the calculated dynamics from Eq. (6.3). Additionally, the experimental results are compared with the Rayleigh equation Eq. (6.1). As indicated by the dashed line in Fig. 6.3b, the 3D bubble model predicts a much faster collapse than what is observed in the experiments.

Here, viscosity is neglected. The time for the diffusive growth of the viscous boundary layer can be expressed as

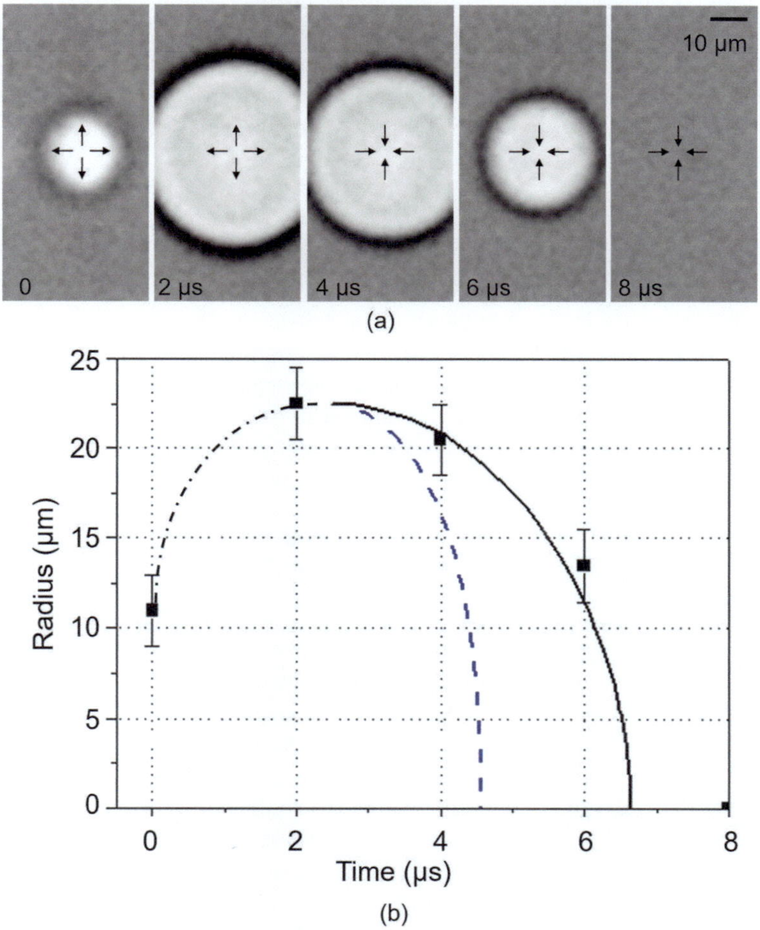

Fig. 6.3 **a** Photographs of mcirobubble evolution in a confined channel. **b** Comparison of the measured bubble radii (squares) with model predictions for the 3D collapse (dashed line) and the 2D inviscid collapse (solid line). The dash-dot line indicates the bubble expansion trend

$$t = y^2/\nu, \tag{6.4}$$

where y is the distance that the vorticity can grow from the wall into the channel, and v is the kinematic viscosity of the liquid. It roughly takes 16 μs for the boundary layer to grow to the center of an 8-μm height channel with $v = 10^{-6}$ m²/s. In Fig. 6.3, the bubble lifetime is approximately 7 μs, thus y is estimated to be 2.6 μm, which is far from the center of the channel. Therefore, the measured bubble dynamics shows good agreement with the inviscid model.

However, the dynamics for larger bubbles can no longer be described by a simple inertia-driven model, i.e., Eq. (6.3). Figure 6.4a shows that the good agreement between the 2D inviscid model and experimental results diminishes as the bubble radius increases. This is because a larger bubble has a longer lifetime, allowing more time for the boundary layer to diffuse from the channel wall and affect the inflow. Another reason is that a larger bubble is generated with higher laser energy, which can introduce more noncondensable gas into the bubble. The collapse process of the microbubble can be hindered by the effects of viscosity or increased gas pressure.

Taking viscosity and noncondensable gas pressure into account, 2D viscous Rayleigh-Plesset-type model has been derived as

$$\left(R\ddot{R} + \dot{R}^2 + \frac{12\nu}{h^2}R\dot{R}\right)\ln\frac{R}{R_\infty} + \frac{\dot{R}^2}{2}\left(1 - \frac{R^2}{R_\infty^2}\right) = \frac{p(R_\infty) - p(R)}{\rho}. \tag{6.5}$$

The third term inside the parentheses on the left-hand side arises from drag forces at the channel walls, which become particularly significant in nanochannels. It is worth noting that the viscous Rayleigh-Plesset-type equation approaches the inviscid version as the channel height approaches infinity ($h \to \infty$). In such cases, the normal stress associated with liquid viscosity may be negligible compared to the channel wall stress. The bubble wall pressure is expressed as

$$p(R) = p_v + p_g - \Delta p = p_v + p_{g0}\left(\frac{R_0}{R}\right)^{2\kappa} - \sigma\left(\frac{1}{R} + \frac{2\cos\theta}{h}\right), \tag{6.6}$$

where p_v is the vapor pressure, p_g is the partial pressure of noncondensable gases with polytropic index κ, p_{g0} is the initial partial pressure of noncondensable gases when the bubble radius is R_0, and Δp denotes the Laplace pressure that depends on the surface tension σ and the contact angle θ. It is assumed that the mass of gases in the bubble remains unchanged.

Figure 6.4b depicts the changes in microbubble radius over time when the bubble content consists solely of vapor. The simulation is based on Eq. (6.5), with the bubble pressure $p(R)$ set as zero. The collapse rate decreases as viscosity increases. When the viscosity is raised from 0 to 1.5×10^{-6} m²/s, the collapse time extends from 7.8 to 12.7 μs, indicating that viscous forces hinder bubble collapse. The collapse dynamics can also be modified by introducing noncondensable gases into the microbubble. Figure 6.4c shows the effect of including polytropic gases ($\kappa = 1.0$) within the

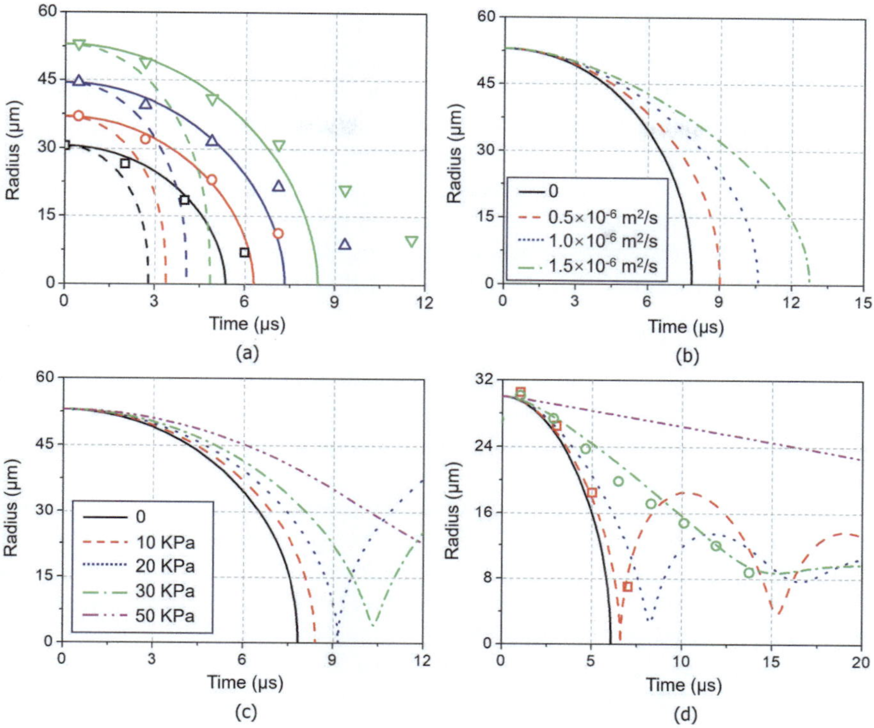

Fig. 6.4 **a** Comparison of bubble dynamics for different-sized bubbles using a 2D inviscid model (solid line) and a 3D Rayleigh model (dashed line). Simulation results for a 2D viscous Rayleigh-Plesset-type model with varying **b** viscosities and **c** gas pressures in a microchannel ($h = 8$ μm). **d** Bubble dynamics for the 2D viscous Rayleigh-Plesset-type model with simulation results in different channel heights: 10 μm (solid), 8 μm (dashed), 6 μm (dotted), 4 μm (dash-dot), and 2 μm (dash-dot-dot), and experimental results in 8 μm (squares) and 4 μm (circles) channels

microbubble, while viscosity is disregarded. When the initial partial pressure of gases (p_{g0}) is quite low, i.e., $p_{g0} < 5$ kPa, the bubble collapse is nearly identical to that of a pure vapor bubble. As the gas amount increases, the overall decay rate is slowed due to gas compression. Simulations reveal an oscillation following the initial collapse when p_{g0} exceeds 20 kPa (see Fig. 6.4c). However, this oscillation is not observed in the experiments, as shown in Fig. 6.4d.

Figure 6.4d shows the bubble collapse with different channel heights. In the simulation, the viscosity is set at 1×10^{-6} m^2/s and the initial gas pressure at 15 kPa. As the channel height decreases, the bubble collapse time is extended. For example, a bubble with a radius of 30 μm in a 10-μm microchannel has a collapse time of 6 μs, while in a 4-μm microchannel, the collapse time exceeds 15 μs. Additionally, bubble oscillation occurs with noncondensable gas, which is particularly evident in the 8-μm and 6-μm microchannels. Oscillation is not observed in the 10-μm channel due to the larger bubble volume, which causes the gas pressure to be negligible. The oscillation is gradually dampened with decreasing channel height, as shown in the

2-μm channel in Fig. 6.4d. The experimental results in the 8-μm and 4-μm channels exhibit the same collapse tendency as the simulation results, though no oscillation is observed. The oscillation is likely influenced by the drag force due to wall shear stress. In smaller scale confinements, such as nanochannels, surface forces are expected to play a more important role in the bubble dynamics.

In addition, the simulations considered different contact angles, using 0° to represent hydrophilic surfaces and 90° for hydrophobic surfaces. However, the results showed no difference in bubble dynamics. Therefore, the effect of surface hydrophobicity on microbubble dynamics can be disregarded.

Dynamics of Nanobubbles

Nanobubble dynamics are studied in nanochannels with heights ranging from 300 to 900 nm. The bubble is generated in the center of a nanochannel with a length of 3 mm and width of 1 mm. The discussions focus on the impact of channel height and surface wetting conditions.

Figure 6.5 compares the dynamics of bubbles generated in channels with heights of 8 μm and 700 nm, both at the same laser energy in the lateral center of the channel. Linear absorption of the laser energy leads to the explosive vaporization of the liquid and rapid expansion of a bubble centered around the focal spot. The bubble is constrained into an approximately cylindrical shape by the channel's top and bottom surfaces. Although the laser energy for both cases is very similar, a much smaller bubble radius is obtained in the nanochannel compared to the microchannel. This is likely due to the significantly less liquid available for vaporization in the nanochannel and the different viscous pressure heads in both channels. While the expansion phase is resolved with two frames for the microchannel, the bubble reaches its maximum size in less than 0.4 μs. The comparable collapse times, despite the different maximum bubble radii, indicate that the collapse is strongly hindered by the confinement. Notably, even after the vapor-bubble-type collapse (the so-called Rayleigh collapse), a residual bubble remains for some time (depicted here after 15 ms). As discussed further below, this is a slowly diffusing gas bubble. Due to the small channel height, even minute gas bubbles remain visible.

The effect of viscosity on bubble dynamics can be assessed by comparing the bubble lifetime with the boundary layer growth time. In relatively tall channels, previous studies have ignored the effects of viscosity and surface tension on bubble dynamics. However, based on Eq. (6.4), the boundary layer reaches the center of the nanochannel in less than a microsecond for all nanochannels examined in this work, which is considerably shorter than the bubble lifetime. This allows sufficient time for the boundary layers at the channel walls to fully develop across the channel height.

The assumption of fully developed boundary layers, used in deriving the Rayleigh-Plesset model, should be relaxed to account for the inertial collapse/growth of bubbles during the initial stage when the boundary layer is still developing and viscous effects on bubble dynamics are less significant. To address this, Eq. (6.5) is corrected as

$$\left(R\ddot{R} + \dot{R}^2 + \phi(\delta^*)\frac{12\nu}{h^2}R\dot{R}\right)\ln\frac{R}{R_\infty} + \frac{\dot{R}^2}{2}\left(1 - \frac{R^2}{R_\infty^2}\right) = \frac{p(R_\infty) - p(R)}{\rho}, \quad (6.7)$$

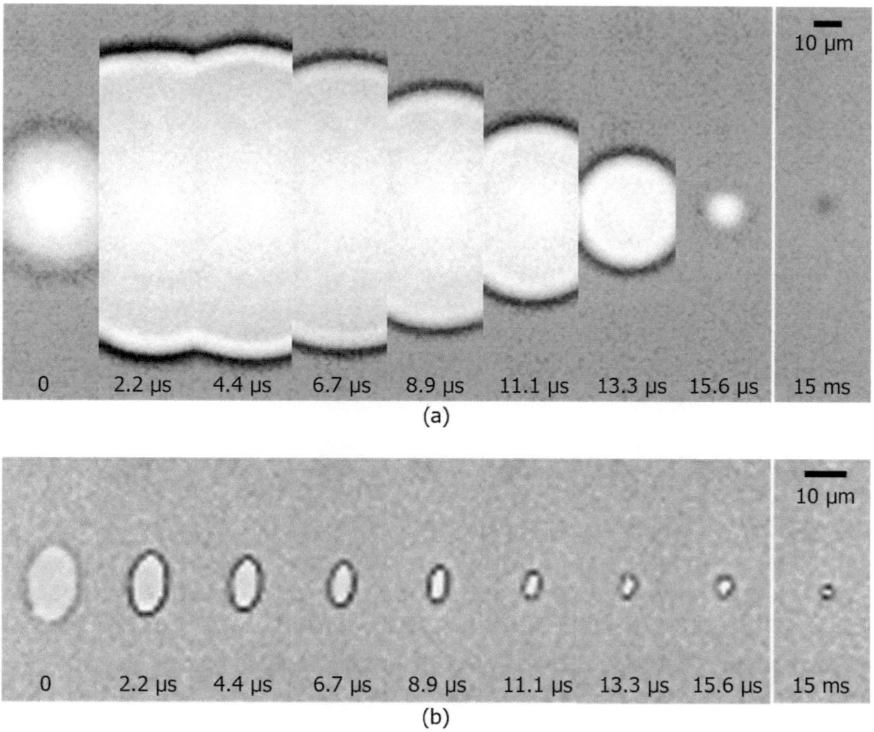

Fig. 6.5 Cavitation bubble evolution in **a** microchannel ($h = 8$ μm) and **b** nanochannel ($h =$ 700 nm)

where ϕ is the correction factor that depends on the boundary layer thickness δ^*, normalized by the channel height. Since the boundary layer thickness at the top and bottom of the channel is approximated by $2\sqrt{\nu t}$, the dimensionless boundary layer thickness may be given by

$$\delta^* = \frac{4\sqrt{\nu t}}{h}. \tag{6.8}$$

To account for the viscous contribution, the correction factor is defined as

$$\phi(\delta^*) = \begin{cases} 0\_\_\text{if}\_0 \leq \delta^* < 1, \\ 1\_\_\text{if}\_\delta^* \geq 1. \end{cases} \tag{6.9}$$

The pressure at the bubble wall in the liquid still can be described by Eq. (6.6).

While the collapse time of vapor bubbles in nanochannels is typically on the order of milliseconds, the overall collapse time in Fig. 6.5 is on the order of seconds. The discrepancy suggests that the bubble also contains noncondensable gases, whose mass diffusion may be sufficiently slow to account for the experimental observations. The classical analysis of the dissolution of a spherical bubble by Epstein and Plesset [25] is extended to a cylindrical gas bubble. Due to the Laplace pressure, the gas pressure is higher than the partial pressure in the liquid, as described by Henry's law. The diffusive mass flux follows Fick's law, with gas pressure approximated by $p_g \approx p_0 + \Delta p$, where Δp is the Laplace pressure and p_0 is the ambient pressure. The evolution of bubble dissolution can be expressed as

$$\dot{R} = \frac{4D}{\pi^2} \frac{c_i - c_s(p_g)}{\rho_g(p_g)R} \int_0^\infty \frac{\exp(-D\xi^2 t)}{J_0^2(\xi R) + Y_0^2(\xi R)} \frac{d\xi}{\xi}, \tag{6.10}$$

where D is the diffusion coefficient, ρ_g is the gas density, J_0 and Y_0 are the Bessel functions of the first and second kinds, c_i and c_s are the mass initial and saturation concentration of the gas, respectively.

Figure 6.6a demonstrates the importance of the viscosity on the bubble dynamics in a nanochannel. The measured and calculated bubble dynamics, considering viscous effects in Eq. (6.7), show a strong correlation. Conversely, the inviscid model described in Eq. (6.3) predicts a much faster shrinkage in the nanochannel but remains applicable for the microchannel. Additionally, Fig. 6.6b examines the initial collapse of a vapor bubble with an initial radius of approximately 5 μm in nanochannels with heights between 300 and 700 nm. It is evident that the rate of bubble collapse slows in the thinner channel due to increased wall shear stress. This trend is also well predicted by the viscous Rayleigh-Plesset-type model.

Both for the diffusion-dominated shrinkage of the remaining gas bubble and the inertial bubble dynamics, the contact angle θ is crucial. The last term in Eq. (6.6) acts as a constant driving pressure that accelerates the collapse, speeding up the

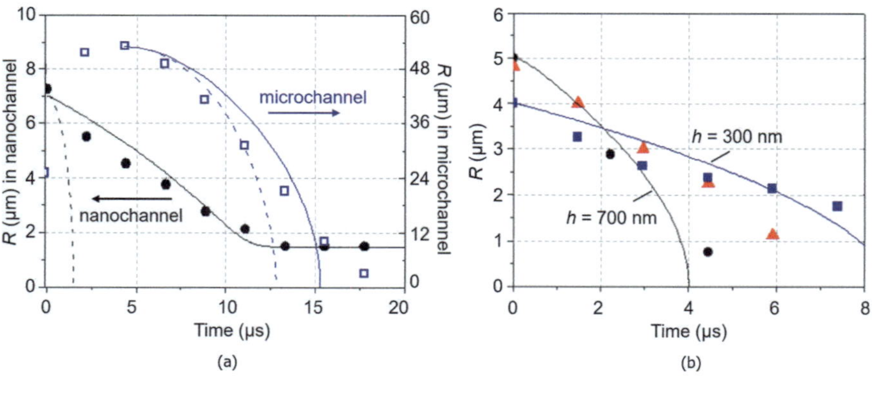

Fig. 6.6 **a** Comparing the radial dynamics of a bubble in nanochannel with $h = 700$ nm (circle) and microchannel with $h = 8$ μm (hollow square) with a viscid 2D model (solid line) and inviscid 2D model (dashed line). **b** Dynamics of cavitation bubble in nanochannels of variable heights: 700 nm (circle), 500 nm (triangle) and 300 nm (square)

bubble collapse in hydrophilic channels. To illustrate this feature of nanochannels, the inertial collapse time and the subsequent diffusion-governed shrinkage are compared for hydrophilic and hydrophobic channels (Fig. 6.7a, b).

The optofluidic chips for bubble experiments are made of borosilicate glass, which naturally has a hydrophilic surface. A silanization treatment with 1H,1H,2H,2H-perfluorodecyltrichlorosilane (FDTS, $C_{10}H_4C_{13}F_{17}Si$, 97%) can make the surface hydrophobic [26]. The presence of a long hydrocarbon chain and fluorine atoms in the organosilane compound causes the treated surface to repel water, making it hydrophobic, as shown in Fig. 6.7c, d. Without any surface treatment, the hydrophilic borosilicate glass has a contact angle of 9°. A silanization treatment with FDTS creates a hydrophobic surface with a contact angle of 90°. The FDTS-treated surface is stable and can withstand long durations, high temperatures, and various chemical treatments.

Figure 6.8a illustrates the inertial collapse time of bubbles with different maximum radii in an extended nanochannel with a height h = 800 nm. The slopes for hydrophobic and hydrophilic cases are markedly different; as expected, bubbles

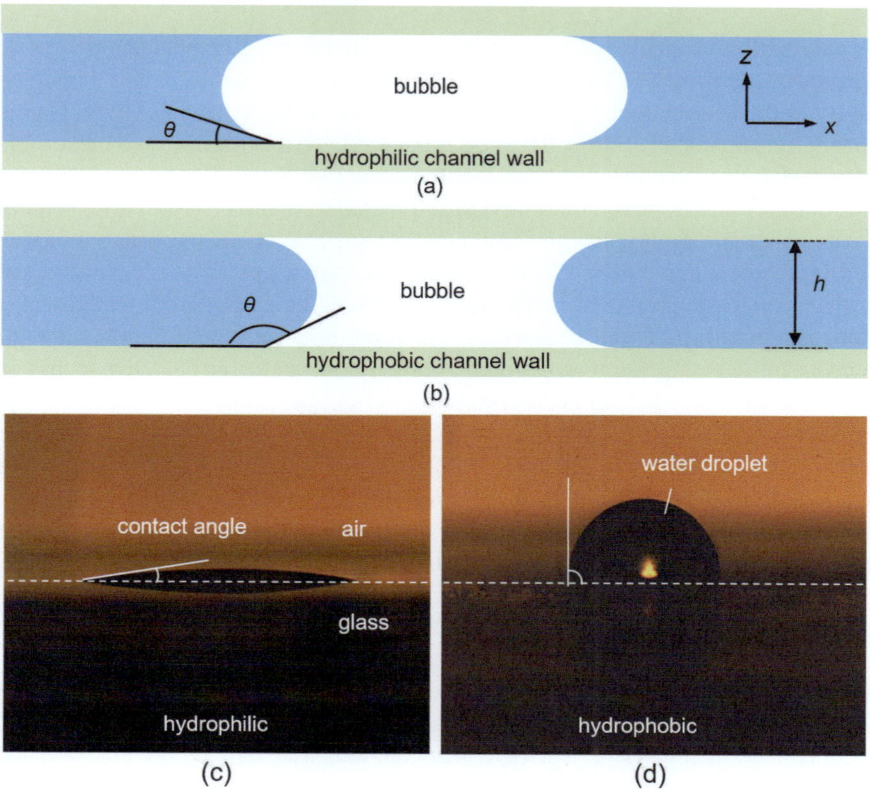

Fig. 6.7 Side view of the nanobubble in the **a** hydrophilic and **b** hydrophobic channels. Photographs of a water droplet on a glass substrate **c** before surface treatment and **d** after surface treatment

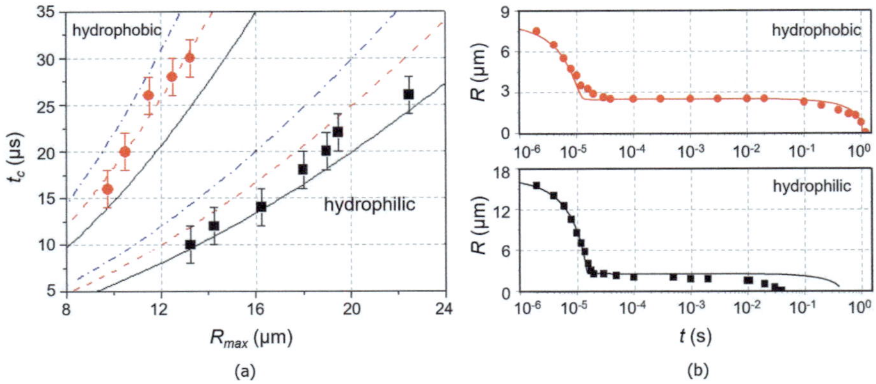

Fig. 6.8 **a** Collapse time as a function of the maximum radius of bubbles in hydrophilic (squares) and hydrophobic (circles) nanochannels. The lines represent the predicted effective viscosities: 0.4 $\times$ 10^{-6} m^2/s (solid line), 0.5 $\times$ 10^{-6} m^2/s (dashed line) and 0.6 $\times$ 10^{-6} m^2/s (dashed-dotted line). **b** Comparison of bubble dynamics in hydrophilic and hydrophobic nanochannels alongside the model predictions

in hydrophilic channels collapse in in a third of the time compared to those in hydrophobic channels (9 μs compared to 29 μs for a 13-μm radius bubble). Interestingly, for the same laser energy, much larger bubbles can be generated in hydrophobic channels compared to hydrophilic ones. It is speculated that during rapid expansion, a three-phase contact line forms, while a thin water film remains on the surface in the hydrophilic case. However, the temporal and spatial resolutions of the current equipment cannot test this hypothesis. The dynamics are also compared with the model, and a reasonable agreement is found. The initial partial gas pressure is unknown, and the viscous Rayleigh model is integrated, assuming an empty void. Additionally, some heating of the liquid is expected due to the laser pulse and illumination. To account for the temperature-dependent viscosity, the results for the range from $\nu = 0.4 \times 10^{-6}$ to 0.6×10^{-6} m^2/s are shown.

Next, the gross bubble dynamics are compared including the slow diffusion of the bubble (as described by Eq. (6.10)). The fast inertial dynamics for the hydrophobic and hydrophilic surfaces agrees nicely with the model, as illustrated in Fig. 6.8b. The experiment reveals that the bubble lifetime of the remaining bubble in the hydrophobic channel is about thirty times longer than that in the hydrophilic channel, i.e., 1.2 s and 40 ms, respectively. Note that the assumption of 2D concentration fields is valid only for cylindrical bubbles ($\theta = 90°$), thus in the hydrophobic channels where a good agreement is obtained. Hydrophilic channels cause a height dependent concentration field due to the small contact angle and can therefore not be described with Eq. (6.10). This may explain the difference between the model and the hydrophilic channel experiments. Here, a static contact angle is used as a representative value in the calculations for the moving contact line, because the (advancing) dynamic contact angle, which is larger than the static one, is unknown in the experiment. The difference between the static and dynamic contact angles (i.e., contact angle hysteresis) depends on the roughness of material surfaces and the velocity of moving contact lines. For bubble dissolution driven by mass diffusion, the bubble collapse rate is sufficiently slow to expect negligible hysteresis effect. However, the vapor collapse rate is rather fast so that such effect may come into play. However, the surface roughness of the channel fabricated through wet etching does not exceed 5 nm, which is much smaller than the channel dimensions. It is thus expected that calculation errors due to the variation of the contact angle are less than or comparable to those from the uncertainty in liquid viscosity. The effect of increasing the contact angle by 10° is examined. While very little effect is seen for the hydrophilic case, the variation gives rise to 30% error in predicting the collapse time of the vapor bubbles for hydrophobic surfaces.

6.3 Jets from Bubble Interaction

A jet is a stream of matter with a roughly columnar shape, spanning from large-scale structures in the universe to subatomic length scales. It has been widely used in manufacturing, agricultural irrigation, and military technology for decades. Recently, small-scale jets have garnered attention for their potential applications in medical diagnostics, DNA sampling, and nuclear fission. In addition to basic studies of jet behavior, numerous works have focused on the jet breakup process and stability analysis.

Jets can originate from a variety of sources, such as a liquid burst from a nozzle, the impact of drops with liquid pools, the impact of a liquid container, and violent eruptions. In this section, the focus is on jets produced by laser-induced bubbles. These jets typically form when a bubble collapses in a liquid with anisotropic pressure. At the end of the collapse stage, the bubble surface generates a rapid liquid jet directed inward against the local pressure gradient. Two widely studied categories are the collapse of a cavitation bubble near a boundary and the compression of a gas bubble caused by a shock wave.

In early studies, liquid jets were observed during the asymmetric collapse of an initially spherical vapor bubble near a rigid boundary. The high-speed liquid jet is directed towards the boundary. Jet formation depends on the movement of the liquid around the bubble. When a cavitation bubble forms near a boundary, the displaced liquid eventually converges at a point, causing the deformation of the bubble surface near that point, which leads to jet formation. The physical properties of the boundary determine the direction of the jet's movement.

Numerous experimental and theoretical studies have been focused on vapor bubble motion near a solid boundary. The bubble migrates towards the boundary, with a jet forming later in the collapse phase. If the bubble is sufficiently close to the boundary, the jet may strike the boundary after piercing the opposing side of the bubble. Various studies estimate the liquid jet velocity during bubble collapse. To directly observe jet development, large bubbles (with a maximum radius of 2 mm) and prolonged collapse times are achieved by reducing the collapse pressure. A maximum velocity of 90–120 m/s is calculated by measuring the tip velocity during the protrusion [27, 28]. Discrepancies between measurements reveal that bubble-boundary interaction during bubble expansion significantly affects subsequent motion. Precise experiments are conducted to study bubble behavior as the distance from the bubble to the solid wall surface and the wall shape are varied.

The behavior of a bubble near a free surface is quite different, even though a jet is also observed during bubble collapse. The free surface acts as a constant-pressure boundary near the pulsating vapor bubble. At larger separations and without buoyancy, the bubble moves away from the free surface during collapse, and the jet pierces the bubble in the direction of bubble migration. When the bubble starts to collapse inward, the free surface develops a pronounced spike moving away from the bubble.

Several studies have investigated flexible boundaries, expecting an intermediate response between the extremes of a rigid boundary and a free surface. Such conditions are often found in biomaterials, where fluidic physics can be very complex. An elastic restoring force can inject momentum back into the liquid, and the boundary's elasticity can be modeled through a pressure term at the interface. The jet direction depends on the elasticity, the bubble's distance from the boundary, and the density ratio. The normalized jet volume is independent of the liquid's density and viscosity but is related to the pressure gradient. These studies are valuable for jets used in biotechnologies. Some bio-applications require a jet directed towards tissue or cells, while others aim to prevent jets from causing damage. Due to their similarity to human tissues and cells, some studies focus on bubble growth and collapse near elastic materials. For instance, the behavior of cavitation bubbles and their interaction with shock waves near a gelatin surface are studied to understand the tissue damage mechanism [29]. Jets with speeds of tens of meters per second are generated, achieving penetration depths of a few millimeters.

In recent years, the bubbles studied have been much smaller than those in earlier research (i.e., several millimeters in radius). It has been reported that bubbles a few microns in size can form a liquid jet in the direction of the propagating shock wave [30]. As the bubble radius increases, the average jet velocity linearly increases from 20 to 150 m/s. At a later stage, the jet breaks up and releases micro-sized bubbles. The jets resulting from the interaction of multiple bubbles have also been investigated. The rapid expansion of a cavitation bubble causes anisotropic compression of an existing stable bubble, forming a penetrating jet. During the expansion stage of the impacted bubble, the nonuniform surrounding flow field induces a transverse inward jetting, which further interacts with the axial jet and leads to fragmentation into small bubbles. Although the bubbles are generated in a 10-μm high liquid gap, their maximum radius reaches 350 μm [31]. Microjets are also observed from the interaction of tandem microbubbles which are generated by laser in a 25-μm liquid layer [32]. Antiphase and coupled oscillation of the tandem microbubbles (with a maximum diameter of 50 μm) lead to the formation of alternating, directional microjets with a maximum velocity of 10 m/s.

The earliest studies on the behavior of jets focused on the formation of drops. Initially, it was believed that a drop falling from a tap was due to gravity. However, the crucial roles of mean curvature were later recognized, revealing that surface tension is the primary force driving the breakup. Surface tension drives the liquid towards a state with a smaller surface area, leading to a decreasing jet radius. Higher surface tension results in faster breakup. In the early nineteenth century, stroboscopic techniques were developed, allowing the recording of jet breakup. It was noted that the breakup occurs spontaneously, independent of any external force or the jet's direction.

The initial growth of the instability is exponential, allowing the breakup length of the jet to be estimated using linear theory [33]. Near breakup, the processes are governed by the balances between surface tension, viscous, and inertial forces. When the jet neck reduces to microscopic scales, thermal noise drives the neck radius to zero. Linear theory for jet breakup starts from random initial perturbations, which are unstable if their wavelength exceeds a critical value. The disturbance amplitude is eventually dominated by the most unstable mode with a wavelength of λ_{opt}. For inviscid jet dynamics, the optimal wavelength is given by $\lambda_{opt} = 9.01\ h_0$, where h_0 is the initial radius of the fluid column [33, 34]. For viscous fluids, jet behavior is influenced by interfacial surface tension and viscous forces. If a long fluid cylinder of viscosity μ' is in an infinite mass of another viscous fluid of viscosity μ, the optimal wavelength λ_{opt} increases as the ratio μ'/μ tends towards infinity or zero [35].

The previously mentioned models do not account for external factors. However, practical experiments are inevitably influenced by environmental conditions, such as turbulence in the bulk liquid or perturbations at the nozzle. As perturbations increase, non-linear effects eventually dominate the breakup process. The dynamics of non-linear jets are explored through increasingly sophisticated experiments, focusing on the formation of satellite drops, minimum thread radius, and drop shape.

Liquid jets are often associated with atomization. During the interpenetration process between two phases, strong shear forces destabilize the central liquid jet, causing it to fragment into a uniform spray. The development of capillary instabilities can be influenced by the extensional motion of the jet. For instance, stretching acts as a damping mechanism for a viscous elongating cylinder [36]. Additionally, the jet can also undergo axial compression, rapidly expanding the jet's cross-section. This expansion forms a radial sheet with a finite extent due to surface tension. The rim of the thin sheet will break up into small droplets, which will be discussed in the following parts.

Most research focuses on liquid jets forming in a gaseous environment and their breakup process. However, the same concepts and similar phenomena should apply to the opposite situation, where hollow (or gas) ligaments form in a continuous liquid phase. Fundamental instability affects elongated void structures. For example, the rippling of the air cavity entrained by a rapidly moving solid object in liquid has been experimentally observed [37]. These cavity ripples induce a spectacularly periodic breakup. Breaking waves in the open ocean entrain air bubbles, and the mechanisms and size distribution of bubble formation are studied [38]. For larger bubbles (>1 mm), turbulent fragmentation determines bubble size, while smaller bubbles are created by jet and drop impacts on the wave face [39].

Recent advancements in optofluidic technology have enabled the study of jet dynamics and stability at micrometer and nanometer scales. An efficient method to create such jets is through laser-induced bubble interactions [40]. In a liquid layer with micrometer thickness, more complex interactions between two bubbles are observed, involving jetting, fragmentation, and entanglement. A thorough understanding of these dynamics and precise control of all parameters remain significant challenges, particularly at small scales. Moreover, the rich phenomena in the transient jetting process motivate further study to uncover the fundamental mechanisms and explore new applications of microjets.

Microjet Processing

A microjet is generated when a static gas bubble is impacted by a laser-induced cavitation bubble in a microchannel as shown in Fig. 6.9. In this example, a microjet is generated in a microchannel with a height of 4 μm and a width of 140 μm. The fabrication processes and experimental setup are the same as discussed in Sect. 6.2. The rapid expansion of the laser bubble generates a shock wave, which compresses the target bubble, resulting in a penetrating microjet. High shear stress causes the gas bubble to rupture during jetting in the confined environment. The thin layer subsequently breaks into several small bubbles, known as shattered bubbles.

The strength of the jet is found to depend on the amplitude of the shock wave, which decreases inversely with the distance (D) between the two bubbles and is proportional to the square root of the laser energy (E). Reflections of the pressure

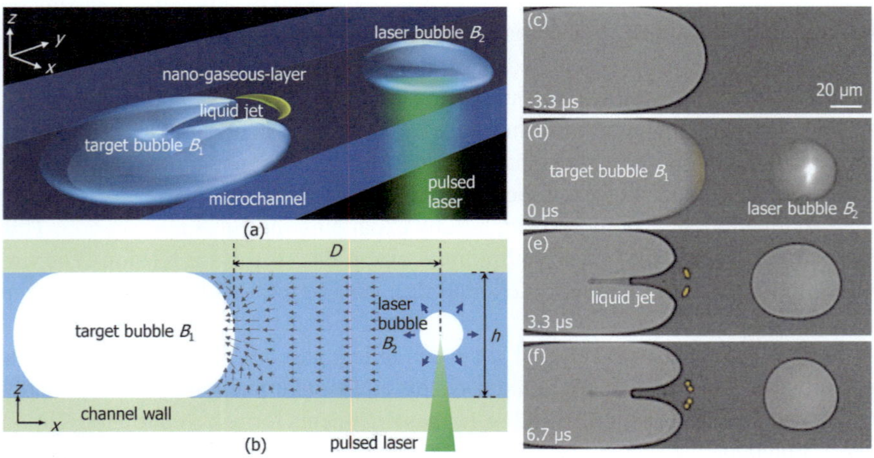

Fig. 6.9 **a** Schematic illustration of microjet formation induced by the interaction between two bubbles in the microchannel. **b** Sideview of the microchannel. **c-f** Sequential snapshots display the process of the target bubble being penetrated by the microjet resulting from the interaction between two bubbles. The evolution of the nano-gaseous layer is highlighted in yellow. Reproduced from [40] with permission from Wiley-VCH

wave at the gas/liquid interface create a pressure gradient (ΔP) between the cavitation bubble and the surface of the gas bubble. When the liquid is set in motion, the curved surface converges with the liquid toward the center of curvature, focusing the flow along the bubble's axial direction in the x-direction. In cases where jetting is caused by the shock wave reflection from the gas/liquid interface, the velocity of the liquid jet (U_{j0}) is twice the liquid velocity (U_p) induced by the shock wave. The pressure wave strength can be estimated from the conservation of momentum.

$$\Delta P \approx \frac{1}{2}\rho c U_{jo}, \tag{6.11}$$

where c is the speed of sound in the liquid. When considering only the dependency of the jet velocity on the pressure wave strength, the jet velocity as a function of energy E and travel distance D of the shock wave can be described as

$$U_{j0} \approx C_0 \frac{(E - E_{th})^{\frac{1}{2}}}{D}, \tag{6.12}$$

where C_0 is a constant factor, and E_{th} is the energy threshold.

Once the jet reaches its maximum length, the retraction process begins. While the initial jet penetration is dominated by inertia, the retraction is controlled by viscosity and surface tension forces. The retraction velocity of the jet can be simplified as

$$U_r \approx \frac{\pi h^2 \sigma}{2\mu L W}, \tag{6.13}$$

where L is the jet length, W is the width of the liquid jet and h is the channel height.

Figure 6.10a illustrates the temporal evolution of the jet velocity and jet length, which is the distance between the original position of the bubble surface and the jet tip. Velocity acceleration is not observed in the experiments, indicating that flow focusing has concentrated the liquid particles rather than increasing the velocity. The initial jet velocity depends on the pressure wave strength. The relatively high Reynolds number (~130) suggests that the initial jet penetration process is dominated by inertia. The boundary layer thickness gradually increases as the flow velocity decreases. Using the equation $\delta t \approx y^2/\nu$ for boundary layer diffusion time (δt) and the kinematic viscosity ($\nu = 1 \times 10^{-6}$ m^2/s), the boundary layer (y) is estimated to reach the center of the channel within 2 μs, at which point viscosity and surface tension become significant factors. The jet reaches its maximum length around 3 μs, after which surface tension drives the jet to retract. During the retraction process, the jet velocity increases to 2.5 m/s within 1 μs. The retraction process is mainly controlled by surface tension and viscous forces. The velocity slowly decreases after the first 1 μs, as the surface tension force gradually diminishes when the jet tip becomes smoother and wider. Figure 6.10b shows the jet velocity as a function of laser energy (E) at different distances (D). The strength of the jet impact increases by either increasing the energy or decreasing the distance between the two bubbles.

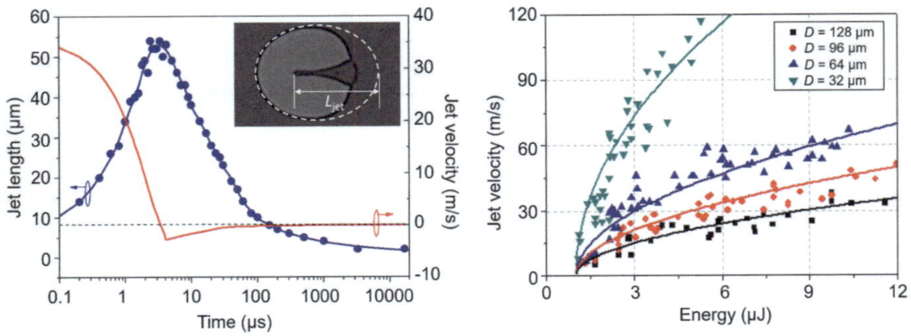

(a) (b)

Fig. 6.10 **a** The evolution of jet length (blue line) and jet velocity (red line). The inset provides the definition of the parameters. **b** Jet velocity as a function of laser energy at various distances

The former increases the pressure and velocity of the laser-induced flow front, as well as the total duration of the expansion stage of the laser bubble, while the latter increases the initial impact.

Pinch-Off and Breakup of Nano-Gaseous-Layers

In a shallow channel, the displacement of the gas/liquid interface varies in the vertical direction due to the high shear stress near the channel wall. Only a small fraction of the liquid that receives the initial impulse and focuses into the jet is in contact with the wall, and therefore its motion is not significantly affected by the pressure wave. The velocity profile in the channel is discussed here.

In the case of rectilinear flow along the x axis, the y and z components of the equation of motion are satisfied by the pressure distribution as

$$p = -\chi(t)x + \rho(g_y y + g_z z) + p_0, \tag{6.14}$$

where $-\chi = \partial p/\partial x$ is the streamwise pressure gradient. The x component of the equation of motion provides an unsteady conduction equation for the streamwise velocity component u_x,

$$\rho\frac{\partial u_x}{\partial t} = \chi(t) + \mu\left(\frac{\partial^2 u_x}{\partial y^2} + \frac{\partial^2 u_x}{\partial z^2}\right) + \rho g_x. \tag{6.15}$$

Time-dependent pressure-driven (Poiseuille) flow is in a channel confined between two parallel walls located at $z = 0$ and $z = h$. The unidirectional flow is due to the application of a constant pressure gradient. After separating variables and using Fourier expansions, the velocity profile is obtained as [41]

$$u_x(z,t) = \frac{\chi}{2\nu\rho}\left[z(h-z) - \frac{8}{\pi^3}h^2\sum_{n=1,3,\ldots}^{\infty}\frac{1}{n^3}\exp\left(-\frac{n^2\pi^2\nu}{h^2}t\right)\sin\frac{n\pi z}{h}\right]. \tag{6.16}$$

The velocity vanishes at the initial instant, $t = 0$. At long times, it recovers the Hagen-Poiseuille parabolic profile.

From the simulation of a microjet in a microchannel ($h = 4\ \mu$m), the flow velocity is nearly uniform except for a small region near the channel wall. Due to the non-slip boundary condition, the velocity is zero at the channel wall and increases to 90% of the maximum velocity at a distance of 240 nm from the wall. This indicates that the plane pressure wave assumption is valid for most parts of the channel. However, a thin layer near the channel wall exhibits a relatively low velocity, affecting the acceleration of the interface due to the velocity profile. While the gas in the center of the channel moves at the maximum velocity, the gas near the channel wall remains almost static. As a result, a thin layer of gas, a few hundred nanometers thick, is attached to the channel wall.

The deformation of the existing gas bubble is amplified by the Richtmyer Meshkov instability, which occurs when a shock wave interacts with an interface between two different fluids. The fundamental mechanism for the amplification of perturbations at the interface is baroclinic vorticity generation, resulting from the misalignment of the pressure gradient of the shock and the local density gradient across the interface. Consequently, a thin layer of gas attached to the channel wall is pinched off from the main part of the gas bubble.

When surface tension transforms the remaining gas layer into a thin cylinder, it subsequently breaks up into small bubbles due to Rayleigh-Plateau instability. Figure 6.11 shows the cylindrical column composed of a fluid with density ρ, bound by surface tension σ. The stability of the column is disrupted by random perturbations, which manifest as sinusoidal components. When the effect of the surrounding fluid is neglected, the wavelength of the varicosity corresponding to the mode of maximum instability is found to be [33]

$$\lambda = 4.51 \times 2a, \tag{6.17}$$

where a is the radius of the cylinder. However, the viscosity (μ) is not negligible in practical situations. The classical stability analysis of Tomotika [36] for a viscous cylinder jet predicts that the unstable mode with the largest growth rate corresponds to

$$\frac{2\pi a}{\lambda} = 0.47, \tag{6.18}$$

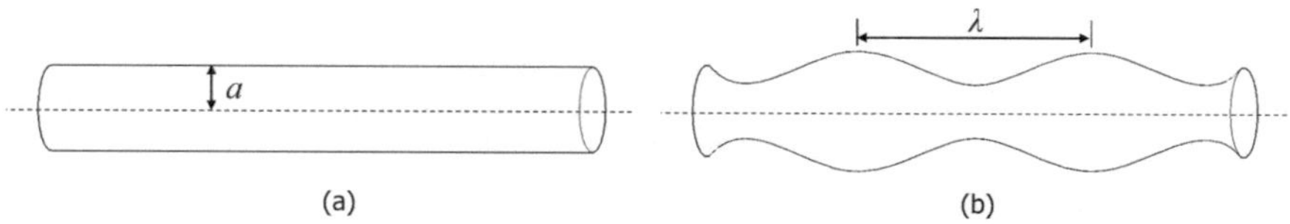

Fig. 6.11 Schematics of the cylindrical column in **a** steady state and **b** perturbed state

for $\mu_i/\mu_o = 0.02$, assuming the cylinder is air ($\mu_i = 1.78 \times 10^{-5}$ Pa s) and surrounded by water ($\mu_o = 1 \times 10^{-3}$ Pa s). The characteristic breakup timescale can be estimated as [42]

$$t_{\text{breakup}} \simeq \sqrt{\frac{\rho a^3}{\sigma}}, \tag{6.19}$$

where ρ is the density of the cylinder and σ is the surface tension. When $a = 1\ \mu\text{m}$, t_{breakup} is 120 ns, meaning that the breakup process occurs too quickly to be captured by most high-speed cameras.

Shattered Bubbles

The most fascinating aspect of microjet is the observation of shattered bubbles. This phenomenon hasn't been reported in other bubble interactions in bulk liquid or channels with larger diameters. Figure 6.12 illustrates the typical sequence of snapshots capturing the fragmentation of the front part of the target bubble. The pressure wave generated by the laser bubble pushes the target bubble surface towards its center. As shown in the 0.3 μs frame in Fig. 6.12, a thin layer of gas is attached to the channel wall due to shear stress. This thin layer is pinched off from the target bubble and constricted into a curved gas cylinder from 0.6 to 1.2 μs. Around 1.5 μs, tiny perturbations cause instability waves to appear. As the instability evolves, the neck between each segment breaks up, shattering the gas cylinder into small bubbles, as depicted in frames from 2.1 to 6.9 μs. In this process, the breakup caused by classical hydrodynamic instability competes with a shrinkage mechanism aimed at reducing surface energy. The breakup of the curved cylindrical gas differs from that of a falling stream, which can be considered infinitely long, or a liquid toroid, where the wavelength of instability must match the tube's circumference. The curved gas cylinder is influenced by its connection points to the main part of gas bubble. In the 1.2 μs frame, the cylindrical gas is seen to separate into four segments. However, the segment number is reduced to two due to boundary conditions influenced by the ends of the curved cylinder. Although the wavelength of instability depends on the cylinder's radius, it can be altered by environmental perturbations, such as the nozzle of jets.

The two segments eventually divide into four small bubbles. This occurs because two layers of remaining gas are attached to the top and bottom channel walls, respectively, and each layer is divided into two bubbles, as shown in the 3.6 μs frame in Fig. 6.12. If the two layers do not break up, each cylinder shrinks into one bubble, resulting in a total of two shattered bubbles, as depicted in Fig. 6.13a. If one layer of gas breaks into two bubbles while the other layer forms a single bubble, the final bubble count is three, as shown in Fig. 6.13b. Based on the double-layer assumption, the thickness of each gas layer can be estimated by assuming the volume of the shattered bubbles is the same as the ruptured gas. The average thickness of the gas layer is 200 nm at 0.3 μs. This agrees with the measured value, determined by the relationship between the thickness of the gas layer and the gray value in the figures.

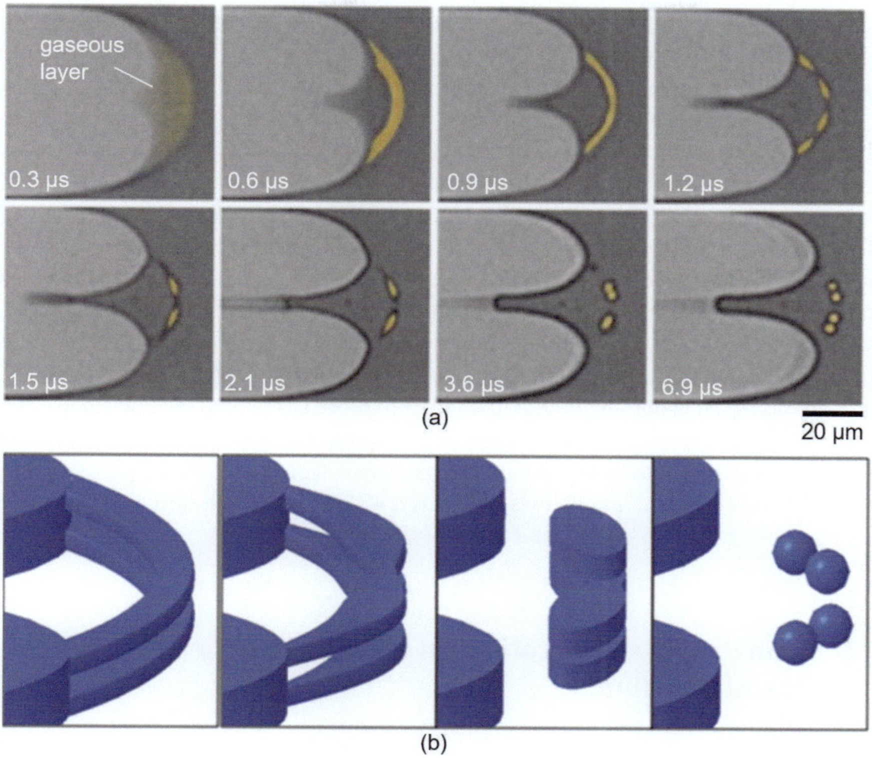

Fig. 6.12 **a** Typical sequential snapshots showing the generation of shattered bubbles during micro-jetting. **b** Schematic illustration of the breakup of nano-gaseous layers. Reproduced from [40] with permission from Wiley-VCH

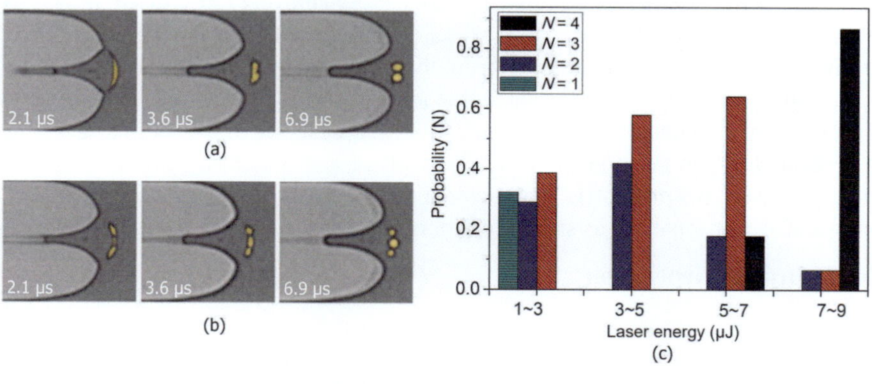

Fig. 6.13 Typical sequential snapshots of the nano-gaseous-layer pinch-off and breakup from the target bubble when the shooting bubble is generated with **a** $E = 4.9 \pm 0.5$ μJ and **b** $E = 6.0 \pm 0.5$ μJ. **c** Histogram of the number of shattered bubbles for different energies with $D = 64 \pm 3$ μm. Reproduced from [40] with permission from Wiley-VCH

The fragmentation number of the shattered bubbles is influenced by laser energy. Figure 6.13c shows the statistical results for the fragmentation number of shattered bubbles with varying laser energy and a fixed distance. When the laser energy is below 1 μJ, shattered bubbles are rarely observed. At lower laser energy, the cylinder's width shrinks slowly, and its length continues to shorten as the connecting points between the main part and the cylinder are pushed toward the axis. For such a short cylinder, only the shrinkage of the cylinder leads to subsequent shape transformation. Even if a bubble remains after the impact, it is quite small and quickly diffuses into the liquid. As energy increases, the probability for a higher number of shattered bubbles increases. In most cases, the pinched-off part will shatter into 2 or 3 bubbles. However, when the laser energy exceeds 7 μJ, four bubbles are predominantly observed. With higher laser energy, the jetting process lasts longer, and the thread is longer when the breakup occurs, causing the thread to break into more bubbles. Based on Eq. (6.18), the estimated wavelength of the varicosity is 13 μm, for a gas cylinder with a radius of 1 μm. This estimation aligns with the measured wavelength of 11 μm. In these experiments, the thread length does not exceed 42 μm, making it difficult to observe shattered patterns with more than four bubbles.

6.4 Nanodroplets Generation via a Single Bubble Transformation

Molecular dynamics simulations indicate that even stable nanometer-sized gas bubbles can jet when impacted by a sufficiently strong shock wave [12]. In confined nanochannels, for timescales shorter than the vortex diffusion time (approximately $t < h^2/v$), inertial-dominated dynamics prevail, which is necessary for flow focusing. Thus, fast jetting may occur within 0.3 μs for $h = 550$ nm and $v = 10^{-6}$ m^2 /s [43]. A laser can be focused very close to a gas/liquid interface in the nanochannel, as shown in Fig. 6.14a. The pressure transient from the explosive bubble induced by the laser pulse generates a nanojet. At the end of the jetting process, the nanojet contracts into a droplet and remains in the gas phase. From the observer's perspective, a bubble is generated by the pulse laser in the liquid phase and transforms into a droplet in the gas phase. This process is analogous to a systematic inverse of a binary object, termed as transformation, as shown in Fig. 6.14b.

Single Droplet Generation

Figure 6.15a–f illustrates the liquid jetting into the gas phase for a stand-off distance of 10 μm. At $t = 0$, the gas/liquid interface is rapidly accelerated to the left (within the 370 ns exposure time of the camera), giving a lower estimate of the jet velocity at 30 m/s. A triangular shape forms on the gas side at $t = 2$ μs, with two sides of the triangle created by two connected liquid columns. The dynamics after $t = 2$ μs slow down significantly. The gas/liquid interface moves outward due to the collapse

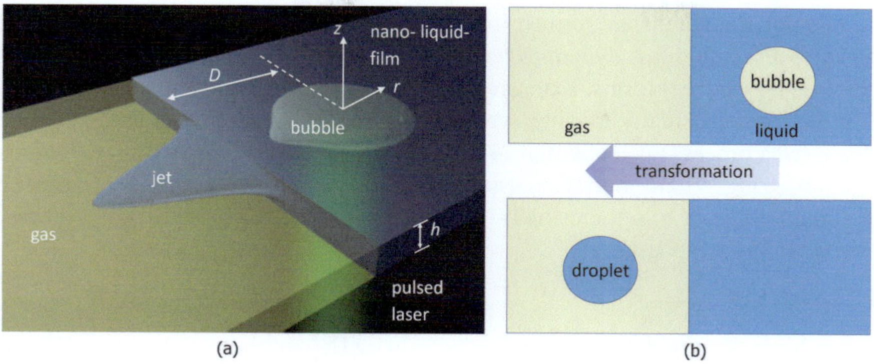

Fig. 6.14 Schematic illustration of **a** jetting caused by a bubble at the gas/liquid interface in the nanofluidic channel and **b** the transformation process of a bubble to a droplet. Reproduced from [43] with permission from the Royal Society of Chemistry

of the nanobubble. Consequently, the liquid columns lose their connection with the external pool and contract into a droplet.

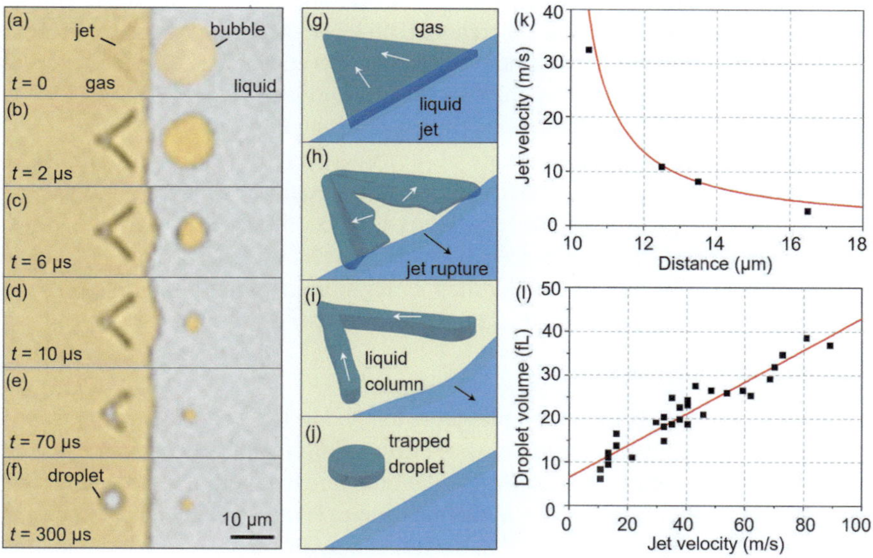

Fig. 6.15 **a–f** Photographs showing the transformation process of a bubble into a droplet in the nanofluidic channel, with the gas phase pseudo-colored in yellow. **g–j** Schematic illustration of the transformation process from a bubble to a droplet. **k** Nanojet velocity as a function of the distance between the laser focus and the bubble surface. **l** Total volume of the droplets as a function of the jet velocity. Reproduced from [43] with permission from the Royal Society of Chemistry

The liquid columns are formed from the breakup of the nanojet. Figure 6.15g–j provides a sketch of the dynamics leading to these structures: Initially, a thin triangular liquid jet sheet forms. Extensional flow, vortices, and/or capillary forces may thin the film, eventually causing it to rupture and leave two liquid columns at its edges. This process occurs within 370 ns and is not resolved due to the limitations of the imaging system. Surface tension further minimizes the surface area, causing the liquid columns to retract slowly. Ultimately, the liquid columns form a droplet trapped in the gas phase.

Jet Dynamics

In Fig. 6.15f, the droplet volume is estimated to be 18 femtoliters using the measured diameter and the channel height of 550 nm. Assuming the conservation of mass, and comparing the jet area in Fig. 6.15a with the droplet volume in Fig. 6.15f, a homogeneous jet with a thickness of 160 ± 30 nm is estimated. This is significantly thinner than the 550-nm confinement. Although the nanojet is milder due to the viscous effect, it provides more precise control over the jetting location at high velocity.

The velocity of the nanojet is highly sensitive to both the laser energy and the distance. Figure 6.15k shows the relationship between nanojet velocity and the distance between the laser focus and bubble surface. The power of the nanojet diminishes more rapidly as the distance increases. The jet velocity drops from 32 to 3 m/s with just a 6 μm reduction in distance. No jet is observed when the distance exceeds 20 μm. Laser energy has a more significant effect at shorter distances. The jet velocity increases proportionally with the laser energy, reaching up to 83 m/s.

As shown in Fig. 6.15l, the volume of the generated droplets is proportional to the jet velocity, as higher jet velocity allows more liquid to penetrate into the gas phase. No droplet forms when the jet velocity is below 5 m/s. However, the droplet size does not always increase linearly with the jet velocity. The volume of a single droplet is found to be always smaller than 25 femtoliters because multiple droplets form when the jet velocity exceeds 20 m/s.

Multiple Droplet Generation

In the discussion above, the thin liquid jet is trapped in the gas phase and contracts into a droplet. Additionally, the thinning of the liquid film can lead to a more complex breakup scenario. Figure 6.16a shows liquid columns breaking up and contracting into three droplets. In some cases, a satellite droplet forms in the center of the film, as shown in Fig. 6.16b. It is likely that the larger opening angle of the triangular jet affects the film drainage, resulting in four distinct and symmetrically placed droplets. When the jet is longer, the liquid columns will break into more droplets. Figure 6.16c provides a clearer illustration of the nanojet breakup due to instability. The thin film breaking pattern is symmetrical. However, the droplet size decreases toward the jet penetration direction, as seen in the third frame of Fig. 6.16c. These droplets develop from liquid plugs on the contour of the original thin liquid film. Interestingly, the plugs can be divided into four parts, with the horizontal distance between each breaking position being the same. Additionally, the curves of the plugs are not continuous.

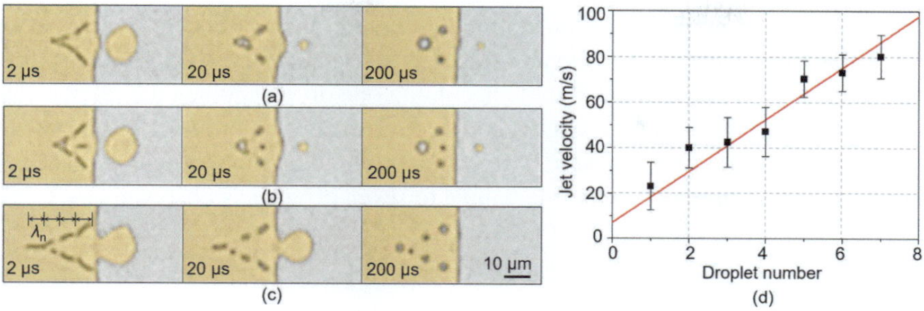

Fig. 6.16 Photographs showing the transformation process of one bubble to multi-droplets: **a** three droplets, **b** four droplets and **c** six droplets. **d** Average jet velocities in the transformation processes with different droplet numbers. The solid lines are linear fittings. Reproduced from [43] with permission from the Royal Society of Chemistry

Based on current evidence, it is reasonable to deduce that drainage is followed by the breakup of the thin film.

The destabilization of the liquid sheet begins with shear instability, also known as Kelvin–Helmholtz instability [44]. When there is velocity shear within a single fluid or a velocity difference across the interface between two fluids, shear instability causes the liquid to move in a flag-like manner. This is followed by Rayleigh–Taylor instability at the sheet's rim, leading to fragmentation. A thin liquid sheet destabilizes when the Weber number (We $= \rho U_0^2 h/\sigma$) exceeds 2 [45], due to shear instability properties with the surrounding medium. From the previous section, the thickness of the liquid sheet (h) is estimated to be 200 nm, and the jet velocity (U_0) is 86 m/s (assuming the jet penetrating time is 300 ns). Thus, the Weber number is approximately 20, indicating that the liquid sheet is unstable. By balancing the destabilizing pressure and the curvature restoration constraint due to capillarity, the wavelength can be expressed as

$$\lambda_0 \approx \frac{\sigma}{\rho_a U_0^2}, \quad (6.20)$$

where ρ_a is the density of the surrounding gas. The measured wavelength λ_n is approximately 6.5 μm, as shown in the first frame of Fig. 6.16c. From Eq. (6.20), the jet velocity should be 96 m/s. This value supports our hypothesis that fast jetting occurs when $t < 300$ ns. The sheet disintegration happens when the amplitude of the undulation continues to increase, causing it to break at the troughs.

Simultaneously, the rim of the liquid sheet recedes due to the Laplace-Young force, which is inversely proportional to the radius of curvature of the rim (~h/2). During this recession, the rim collects liquid. Consequently, the films feeding the rims drain, and the rims progressively contract into liquid plugs. This entire process completes within 2 μs. Then, it takes approximately 200 μs for surface tension to further shape the liquid plugs into spherical droplets.

Generally, the number of droplets increases with the jet velocity, as shown in Fig. 6.16d. For instance, when the jet velocity is between 10 and 30 m/s, the transformation typically results in one bubble becoming one droplet. When the jet velocity reaches 40 m/s, one bubble often turns into three droplets. Sometimes, two plugs fuse during the breakup process, resulting in two droplets. When the jet velocity reaches 80 m/s, the transformation from one bubble to seven droplets is observed.

6.5 Summary

In many practical situations, fluid dynamics is significantly influenced by nearby boundaries. Examples include bubble oscillations in blood vessels and high-speed jetting from nozzles. Researchers are interested in exploring how sub-micrometer confinement affects dynamics. An experimental setup that can be examined using high-speed imaging is an extended nanochannel with one dimension on the nanometer scale, i.e., hundreds of nanometers in height. Nanobubbles are generated within the extended nanochannels filled with light-absorbing liquid by focusing a pulsed laser beam. Previous studies on nanobubbles have rarely used this direct observation method, which also reduces the need for advanced fabrication techniques. A Rayleigh-Plesset-type model is derived to describe the dynamics of a confined bubble. This model considers channel height as a critical factor and includes the effects of surface tension and viscosity. Thus, it is applicable to both microchannels and nanochannels, addressing the limitations of other models that diverge from experimental data as channel dimensions decrease. As the channel height decreases, the bubble lifetime is prolonged.

In micro/nanochannels, a jet is generated when a laser-induced bubble forms near another gas bubble. In a microchannel, the jetting process is long enough to allow for thorough analysis. Additionally, instability causes the formation of shattered bubbles from nano-gaseous layers, which rupture from the gas bubble due to shear stress. In a nanochannel, fast jets are produced within hundreds of nanoseconds, taking the form of a thin sheet approximately 200 nm thick. It has been demonstrated that inertia-dominated dynamics are still possible for sufficiently short times. The transformation process from bubble to droplet has been experimentally shown during the jetting process in the nanochannel. The number and volume of droplets are controlled by the jet velocity, which is proportional to the laser energy and inversely proportional to the distance. This provides new insights into nanodroplet formation, which holds great potential for biological applications such as drug delivery and cell surgery.

References

1. Tsai JH, Lin LW (2002) Active microfluidic mixer and gas bubble filter driven by thermal bubble micropump. Sens Actuators A Phys 97–8:665–671
2. Marmottant P, Hilgenfeldt S (2004) A bubble-driven microfluidic transport element for bioengineering. Proc Natl Acad Sci USA 101(26):9523–9527
3. Marmottant P, Hilgenfeldt S (2003) Controlled vesicle deformation and lysis by single oscillating bubbles. Nature 423(6936):153–156
4. Li ZG, Liu AQ, Klaseboer E, Zhang JB, Ohl CD (2013) Single cell membrane poration by bubble-induced microjets in a microfluidic chip dagger. Lab Chip 13(6):1144–1150
5. Rayleigh L (1917) On the pressure developed in a liquid during the collapse of a spherical cavity. Phil Mag 34(199–04):94–98
6. Lauterborn W, Kurz T (2010) Physics of bubble oscillations. Rep Prog Phys 73(10):106501
7. Parlitz U, Mettin R, Luther S, Akhatov I, Voss M, Lauterborn W (1999) Spatio-temporal dynamics of acoustic cavitation bubble clouds. Philos Trans R Soc A Math Phys Eng Sci 357(1751):313–334
8. Krishna R, Urseanu MI, van Baten JM, Ellenberger J (1999) Rise velocity of a swarm of large gas bubbles in liquids. Chem Eng Sci 54(2):171–183
9. Xiao C, Heyes DM, Powles JG (2002) The collapsing bubble in a liquid by molecular dynamics simulations. Mol Phys 100(21):3451–3468
10. Lugli F, Hofinger S, Zerbetto F (2005) The collapse of nanobubbles in water. J Am Chem Soc 127(22):8020–8021
11. Lugli F, Zerbetto F (2007) Molecular dynamics of nanobubbles' collapse in ionic solutions. ChemPhysChem 8(1):47–49
12. Vedadi M, Choubey A, Nomura K, Kalia RK, Nakano A, Vashishta P, van Duin ACT (2010) Structure and dynamics of shock-induced nanobubble collapse in water. Phys Rev Lett 105(1):014503
13. Parker JL, Claesson PM, Attard P (1994) Bubbles, cavities, and the long-ranged attraction between hydrophobic surfaces. J Phys Chem 98(34):8468–8480
14. Ducker WA (2009) Contact angle and stability of interfacial nanobubbles. Langmuir 25(16):8907–8910
15. Brenner MP, Lohse D (2008) Dynamic equilibrium mechanism for surface nanobubble stabilization. Phys Rev Lett 101(21):214505
16. Noack J, Hammer DX, Noojin GD, Rockwell BA, Vogel A (1998) Influence of pulse duration on mechanical effects after laser-induced breakdown in water. J Appl Phys 83(12):7488–7495
17. Schaffer CB, Nishimura N, Glezer EN, Kim AMT, Mazur E (2002) Dynamics of femtosecond laser-induced breakdown in water from femtoseconds to microseconds. Opt Express 10(3):196–203
18. Vogel A, Linz N, Freidank S, Paltauf G (2008) Femtosecond-laser-induced nanocavitation in water: Implications for optical breakdown threshold and cell surgery. Phys Rev Lett 100(3):038102
19. Kotaidis V, Plech A (2005) Cavitation dynamics on the nanoscale. Appl Phys Lett 87(21):213102
20. Lukianova-Hleb EY, Ren XY, Zasadzinski JA, Wu XW, Lapotko DO (2012) Plasmonic nanobubbles enhance efficacy and selectivity of chemotherapy against drug-resistant cancer cells. Adv Mater 24(28):3831–3837
21. Lapotko D (2009) Plasmonic nanoparticle-generated photothermal bubbles and their biomedical applications. Nanomedicine 4(7):813–845
22. Duan C, Karnik R, Lu MC, Majumdar A (2012) Evaporation-induced cavitation in nanofluidic channels. Proc Natl Acad Sci USA 109(10):3688–3693
23. Plesset MS (1949) The dynamics of cavitation bubbles. J Appl Mech Trans Asme 16(3):277–282
24. Quinto-Su PA, Lim KY, Ohl CD (2009) Cavitation bubble dynamics in microfluidic gaps of variable height. Phys Rev E 80(4):047301

25. Epstein PS, Plesset MS (1950) On the stability of gas bubbles in liquid-gas solutions. J Chem Phys 18(11):1505–1509
26. Shui LL, van den Berg A, Eijkel JCT (2009) Interfacial tension controlled W/O and O/W 2-phase flows in microchannel. Lab Chip 9(6):795–801
27. Lauterborn W, Bolle H (1975) Experimental investigations of cavitation-bubble collapse in neighborhood of a solid boundary. J Fluid Mech 72:391–399
28. Shima A, Takayama K, Tomita Y, Miura N (1981) An experimental-study on effects of a solid wall on the motion of bubbles and shock-waves in bubble collapse. Acustica 48(5):293–301
29. Kodama T, Tomita Y (2000) Cavitation bubble behavior and bubble-shock wave interaction near a gelatin surface as a study of in vivo bubble dynamics. Appl Phys B Lasers Opt 70(1):139–149
30. Ohl CD, Ikink R (2003) Shock-wave-induced jetting of micron-size bubbles. Phys Rev Lett 90(21):214502
31. Chen YH, Lin I (2008) Dynamics of impacting a bubble by another pulsed-laser-induced bubble: jetting, fragmentation, and entanglement. Phys Rev E 77(2):026304
32. Sankin GN, Yuan F, Zhong P (2010) Pulsating tandem microbubble for localized and directional single-cell membrane poration. Phys Rev Lett 105(7):078101
33. Rayleigh L (1878) On the instability of jets. Proc Lond Math Soc 10:4–13
34. Rayleigh L (1892) On the instability of a cylinder of viscous liquid under capillary force. Phil Mag 34(207):145–154
35. Tomotika S (1935) On the instability of a cylindrical thread of a viscous liquid surrounded by another viscous fluid. Proc R Soc Lond Ser A Math Phys Sci 150(870):322–337
36. Tomotika S (1936) Breaking up of a drop of viscous liquid immersed in another viscous fluid which is extending at a uniform rate. Proc R Soc Lond Ser A Math Phys Sci 153(A879):0302–0318
37. Grumstrup T, Keller JB, Belmonte A (2007) Cavity ripples observed during the impact of solid objects into liquids. Phys Rev Lett 99(11):114502
38. Loewen MR, Odor MA, Skafel MG (1996) Bubbles entrained by mechanically generated breaking waves. J Geophys Res Oceans 101(C9):20759–20769
39. Deane GB, Stokes MD (2002) Scale dependence of bubble creation mechanisms in breaking waves. Nature 418(6900):839–844
40. Xiong S, Chin LK, Tandiono T, Liu A-Q, Ohl C-D (2014) Microjet-initiated nano-gaseous layer pinch-off from the surface of a bubble and subsequent breakup. Isr J Chem 54:1602–1606
41. Pozrikidis C (2011) Introduction to theoretical and computational fluid dynamics, 2nd edn. Oxford University Press, New York
42. Eggers J (1997) Nonlinear dynamics and breakup of free-surface flows. Rev Mod Phys 69(3):865–929
43. Xiong S, Chin LK, Ando K, Tandiono T, Liu A-Q, Ohl C-D (2015) Droplet generation via a single bubble transformation in a nanofluidic channel. Lab Chip 15:1451–1457
44. Thomson W (1871) Hydrokinetic solutions and observations. Phil Mag 42(281):362–377
45. Villermaux E, Clanet C (2002) Life of a flapping liquid sheet. J Fluid Mech 462:341–363

Chapter 7
Optofluidic Droplet Fusion

7.1 Introduction

The process of droplet fusion has been studied for centuries, with contributions from various fields such as physics, chemistry, and engineering. This process is vital for numerous industrial applications, including printing and sintering, as well as chemical and biological processes like emulsification and extraction. Recent advancements in droplet microfluidics research have highlighted the need to further study this phenomenon [1–3]. Microfluidic devices allowed for precise control over droplet generation, manipulation and fusion, leading to a better understanding of the underlying mechanisms. Droplet fusion has found applications in diverse fields such as materials synthesis, single-cell analysis, RNA sequencing, small molecule screening, in vitro diagnostics, and tissue engineering [4]. Today, research continues to explore new methods and applications of droplet fusion, with a focus on improving efficiency, scalability, and integration into various systems. The techniques developed for this purpose can be categorized into two groups: passive and active droplet fusion techniques.

Passive techniques for droplet fusion in microfluidic systems have been developed based on channel geometries, partial surface modification, and direct injection using side channels. Various channel geometry designs have been implemented to fuse droplets, such as sudden expansions with rectangular shapes, disk shapes, tapered expansions, and fluid resistance elements. In a rectangular sudden expansion structure within a microfluidic channel, fluids are removed from the expansion channel part in equal volumes through the top and bottom channels [5, 6]. The volume drained between two droplets can be controlled by adjusting the flow rates, enabling control over the number of droplets to be fused. A controlled chemical reaction in a droplet microreactor is achieved using a disk-shaped expansion microchannel design [7]. The volumetric ratio between the continuous phase and dispersed phase is managed by adjusting the flow rate ratio before contact. The synthesis of cadmium sulfide (CdS) nanoparticles has been demonstrated in microchannels with a tapered expansion

S. Xiong and A.-Q. Liu, *Optofluidics: Fundamentals and Applications*,
Microfluidics and Nanophotonics: Science and Engineering 4,
https://doi.org/10.1007/978-981-95-6296-1_7

design by fusing two droplets with different reagent solutions [8]. Fluid resistance elements have also been utilized to fuse droplets [9]. As a functional element, the fluid resistor is combined with injectors, segment splitters, and transformation elements to form a complex fluidic microreactor. Generally, these geometric methods leverage the same basic mechanism: the expanded portion in the microchannel thins the continuous phase, i.e., drains the film between droplets, by decreasing the droplets' flow rate, while the film ruptures downstream in the contracting channel.

Another group of passive techniques involves surface modification to change the surface energy [10]. For example, a partially hydrophilic microchannel can be used to trap and fuse droplets. A pair of droplets, one containing distilled water and the other a dye solution, is demonstrated to be trapped by the hydrophilic part of the microchannel, where they are then fused and mixed. The fusion process occurs within tenths of milliseconds to milliseconds. At high flow rates, the droplets will flow directly through the hydrophilic stripes without fusing. The detachment of the fused droplet or the fusion of additional droplets can be controlled by adjusting the flow rate of the continuous phase fluid.

Some research groups use side channels to directly inject additional reagents into target droplets [11]. Different reagents are introduced into passing droplets through a set of three side channels. This method reduces the probability of cross-contamination and prevents the generation of small droplets of additional reagents. The side channel method has also been used to inject a third immiscible liquid as spacers between droplets to prevent their fusion [12].

Passive droplet fusion offers several advantages, including simple design, non-active control, and low contamination risk. However, the fusion rate is inherently limited by the passive fusion mechanism, making it unsuitable for on-demand droplet fusion. With passive methods, any droplet passing through the functional microchannel will fuse. Active fusion methods can overcome this limitation and achieve on-demand droplet fusion.

Several active techniques for droplet fusion using microfluidic systems have been developed and demonstrated using electrical, heat-based and optical methods. For electrical methods, electrodes are usually integrated with microchannels to induce electrical forces on the target droplets, which lead to an imbalance of the interfacial stress forcing the fusion of adjacent droplets. With parallel electrodes oriented perpendicular to the droplet flow direction, the electric field parallel to the flow direction is generated to induce the force to deform two adjacent droplets and the electrocoalescence [13]. The measurement of an enzyme kinetic constant has been demonstrated using this method with high throughput at rates of thousands droplet per second coalescence. The fusion of tightly packed droplets has been demonstrated using a low voltage potential pulse [14]. A pair of gold electrodes is fabricated perpendicular to the microchannel. When a potential pulse is applied, the pair of droplets between the two electrodes coalesce in 100 μs or less. After the optimization of the electrode geometry and the applied potential, the rupture of the single lamellae within highly ordered droplet arrangements has been demonstrated. Droplets and cell electrofusion are also demonstrated using continuous DC voltage [15]. When a pair of linked cells based on biotin-streptavidin interaction flows through the microchannel with geometry variation, the electrofusion is conducted. The geometry of the microchannel effectively avoids the fusion of more than three cells.

Dielectrophoresis (DEP) has also been utilized to fuse droplets [16]. The liquid medium of droplets is dielectrically distinct from the continuous phase liquid. Under non-uniform electric stresses induced by electrodes integrated at the bottom of a microchannel, droplets are deformed and stretched. When two droplets come within a sufficiently close distance, they coalesce. In another approach, a droplet is driven by sequentially turning on and off a series of electrodes, moving it toward another droplet to initiate coalescence [17]. The fusion of aqueous droplets in an immiscible and low-permittivity continuous phase is demonstrated through both moving droplets with an array of two-dimensional electrodes and dielectrically-activated injection of droplets.

Droplet fusion has also been demonstrated using a combination of the sudden expanded channel design and aligned parallel electrodes. With the sudden expanded channel design, stabilized surfactant droplets are brought together, and the fusion process is activated by a pulsed electrical field [18]. Droplet fusions of various sizes, from nanoliters to picoliters, are achieved, with precise control over the start of fusion. Electrical methods have also been applied to the fusion of more stable vesicles, such as cells and liposomes [19]. By using an array of asymmetric electrodes, a large fraction of liposomes are observed to fuse. When DC pulses are applied to the vesicles, small pores form on the membrane of the liposome vesicles. When two vesicles are in close proximity, their porous membranes rearrange and fuse together.

Another set of active methods involves using heating elements combined with a microfluidic resistance element [9]. By increasing the temperature, the viscosity of the fluid decreases, which enhances the drainage rate. When combined with a microfluidic resistance element, the lower viscosity liquid flows with the activation of the heating elements, slowing the movement of the droplets and triggering their fusion.

Optical tweezers are also popular for on-demand droplet fusion through direct optical manipulation [20]. Droplets with a lower refractive index are generated in the immiscible continuous phase. Optical vertex traps are employed to manipulate and transport these droplets, facilitating fusion. However, due to the low forces exerted by optical tweezers, this method is only suitable in a stationary liquid or at very low flow rates.

Alternatively, a laser-based droplet fusion technique is proposed [21]. On-demand fusion of stationary and moving droplets has been achieved by generating a short-lived cavitation bubble within one droplet. This process is completed on an optofluidic chip, designed for microdroplet-based chemical synthesis and bioassays. This chapter explores this optofluidic technique, demonstrating fusion for static and moving droplets, as well as droplets of equal and unequal sizes. Furthermore, it reveals diffusion-dominated mixing flow and the transport of a single encapsulated cell into a fused droplet.

7.2 Mechanism of Droplet Fusion

Typically, the fusion process of two droplets can be divided into three consecutive steps. The first step involves the film drainage of the continuous phase, followed by the film rupture in the second step. The third step is the growth of the connecting neck between the two merging droplets. In this chapter, rapid thinning of the continuous phase film during bubble growth is achieved by creating a cavitation bubble inside one droplet near another. The film rupture is observed during the shrinkage of the bubble, while the growth of the connecting neck occurs on a much longer timescale after the bubble has disappeared.

These three stages of laser-induced droplet fusion are shown in Fig. 7.1. Droplet A and Droplet B are two equal-sized droplets positioned close to each other (Fig. 7.1a). A laser-induced cavitation bubble is created inside one of the two droplets (droplet A) (Fig. 7.1b). The center of the cavitation bubble is focused at the center of droplet A. The expanding bubble contains vapor of the dispersed phase, which "blows up" and pushes droplet A into the neighboring droplet B. The expansion of droplet A is not isotropic but more pronounced towards droplet B. This is due to the higher viscosity of the surrounding fluid (e.g., oil), leading to a higher normal stress difference sideways across the interface. In contrast, the effective viscosity of the fluid in droplet B (e.g., water) is smaller. Therefore, droplet A primarily deforms towards droplet B, causing droplet B to take on a heart shape. The initial continuous film is rapidly thinned by the impulsive force of the expanding bubble. After the bubble reaches its maximum volume, a pressure imbalance causes the bubble to reverse its motion, leading to contraction and subsequent collapse. This pulls the thin film between the two droplets back towards droplet A (Fig. 7.1c). It is speculated that the reversal of motion, i.e., the acceleration, causes the film to rupture. Following this, a neck connecting the two droplets becomes visible, and the fusion process occurs at a much slower rate, forming a larger droplet (Fig. 7.1d–f). Due to interfacial tension, the surface energy between the two liquids is minimized as the neck diameter increases. Ultimately, a single convex-shaped droplet is formed, which eventually becomes spherical with minimal surface energy (Fig. 7.1f).

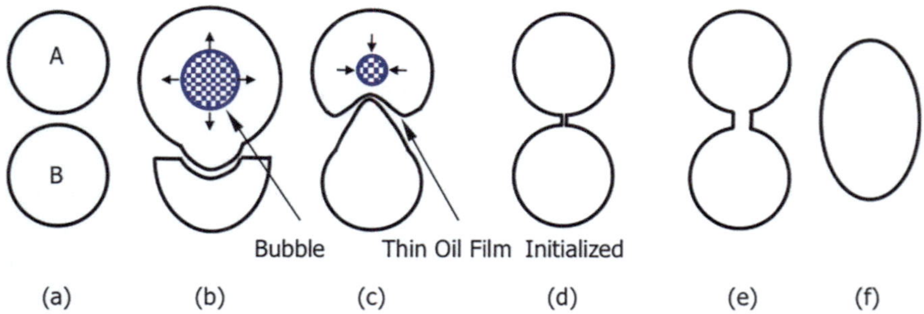

Fig. 7.1 Schematic process of fusing two equal-sized droplets using a laser-induced cavitation bubble. Reproduced from [21] with permission from the Royal Society of Chemistry

During the droplet fusion process, when two droplets come close, the thin film of the continuous phase between them drains away due to the applied gravitational or inertial force. Electrostatic repulsive forces can slow the drainage rate as the film thins. When the film becomes very thin, the drainage rate significantly decreases. London-van der Waals attractive forces cause the film to rupture, and a spontaneous connection, mediated by capillary waves, occurs faster than further drainage, leading to the fusion of the two droplets.

After the film ruptures, the fusion process initiates with the formation and growth of the neck between the two droplets. The growth speed of the neck is reduced due to the interplay between capillary and viscous forces. While the capillary force drives the fusion process, the viscous forces slow it down. By formulating these two forces (i.e., normalizing the capillary number to unity), the time dependence of the neck radius R_t can be expressed as [22]

$$R_{\mathrm{t}}(t) \propto \frac{\gamma}{\eta} t \tag{7.1}$$

where γ is the surface tension, η is the viscosity, and t is the time. This fusion mechanism results in high speeds for ordinary liquids, with the capillary velocity of water being approximately 70 m/s.

The theory predicts a logarithmic correction to the above scaling argument [22], and the modified Eq. (7.1) is given as follows,

$$R_{\mathrm{t}}(t) = -\frac{\gamma}{\pi \eta} t \ln\left(\frac{\gamma}{R_0 \eta} t\right), \tag{7.2}$$

where R_0 is the initial radius of the undeformed droplet. This prediction is valid for viscous droplets in an inviscid surrounding.

The primary force driving the fusion dynamics can be either viscous or inertial, determined by the ratio of inertial forces to viscous forces, known as the Reynolds number (Re) (refer to Eq. (2.10)). If Re > 1, inertial fusion occurs. If Re < 1, viscous fusion takes place.

7.3 Fusion of Two Equal-Sized Droplets

Based on theoretical analysis, a microfluidic chip with a T-junction is designed and fabricated to generate water-in-oil droplets as described in Sect. 5.2. To produce droplets of various sizes, the flow rates of water and oil are initially set to a suitable value to consistently generate droplets. Then, the water inlet is slowly switched off to create a series of droplets with decreasing diameters. These droplets are collected in a chamber. The experimental setup for creating a laser-induced cavitation bubble inside the droplets is depicted in Chap. 6. The growth process of the connection neck

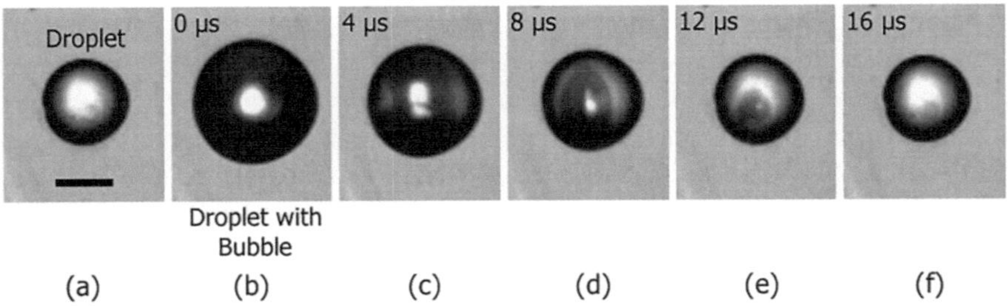

Fig. 7.2 Selected frames showing the dynamics of a single cavitation bubble within an aqueous droplet. The scale bar represents 20 μm

between the droplets is captured using an ultra-high-speed camera. The neck radius as a function of time is measured and analyzed using high-speed micrographs.

Bubble dynamics inside a single water droplet

Initially, the dynamics of a cavitation bubble in a droplet are investigated. Figure 7.2 illustrates the creation of a laser-induced cavitation bubble in an aqueous droplet. The bubble forms a few micrometers above the bottom of the microfluidic channel. When the laser hits the center of the droplet, the bubble is generated due to the absorption of water and dye molecules. High-speed micrographs are recorded at framing rates up to 250,000 frames per second, with an exposure time of 370 ns. The first micrograph showing the bubble is defined as time $t = 0$ (Fig. 7.2a). The bubble quickly forms after the laser pulse, reaching its maximum diameter at $t = 0$ μs, causing the droplet to expand to its maximum size simultaneously (Fig. 7.2b). The bubble then starts to shrink, and by $t = 16$ μs, it completely collapses within the droplet (Fig. 7.2c–f). When the cavitation bubble is created near the lower boundary of the droplet, its dynamics change significantly. Due to the boundary-bubble interaction effect, a jet forms towards the boundary. The bubble then splits into two sub-bubbles and eventually collapses.

The droplet size is measured, and the variation of the diameters over time is plotted in Fig. 7.3. The black squares represent the radius of the droplet, while the red dots indicate the radius of the laser-induced cavitation bubble inside the droplet. Using the spherical volume equation, the bubble radius R_b at different times can be determined by

$$R_b = (R_d - R_{d0})^{1/3}, \tag{7.4}$$

where R_d is the radius of the droplet, and R_{d0} is initial radius of the droplet. This equation is based on two assumptions: (1) both the droplet and the cavitation bubble maintain ideal spherical shapes throughout the experiment, and (2) the volume of the droplet remains constant during the experiment, despite minor mass losses caused by the vaporization of a small portion to generate the bubble.

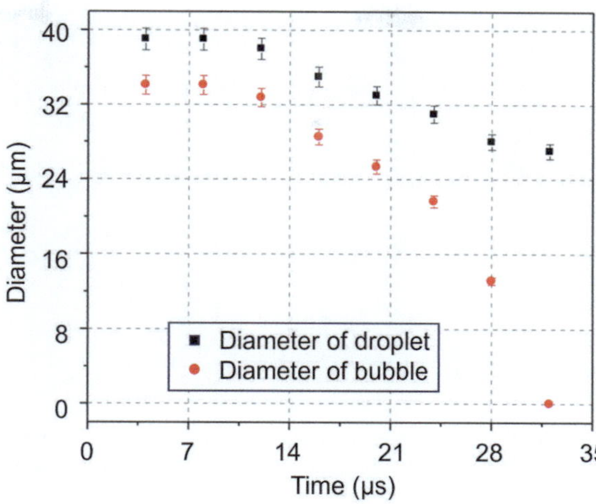

Fig. 7.3 Evolution of the diameters of the droplet and the bubble within the droplet

Equal-sized droplets are generated in a microfluidic channel and then stored in the droplet collection chamber after the flow has been switched off. Experimental micrographs, selected from a high-speed recording of the fusion of two equal-sized droplets in a stagnant liquid, are shown in Fig. 7.4. The sequence spans the time from 4 μs before bubble formation to the fully developed neck at $t = 1$ s. The first micrograph in Fig. 7.4 shows two aqueous droplets separated by a 6-μm liquid oil film. The upper droplet has a diameter of 92 μm, while the lower droplet has a diameter of 96 μm. The laser pulse is focused on the center of the upper droplet, creating a vapor bubble due to the absorption of light in the dispersed phase. The initial hot vapor of the dispersed phase causes the bubble to grow explosively. The expansion of the upper droplet is not isotropic but more pronounced towards the lower droplet. As the bubble rapidly expands, the vapor inside cools down quickly, reducing the internal pressure. This decelerates the radial expansion, and the bubble reaches its maximum size at approximately $t = 0$ μs, within about 4 μs. The bubble begins to shrink and collapses at $t = 32$ μs, leaving a slowly dissolving gas bubble inside the upper droplet. Afterward, the bulged interface of the lower droplet gradually retracts downward. The neck of the unconnected droplets shrinks between $t = 30$ ms and $t =$ 50 ms. It is noteworthy that fusion occurs during the retraction of the interface. The fusion process, characterized by the formation of a clear connecting neck, becomes visible at approximately $t = 65$ ms. The thickness of the neck gradually increases until $t = 1$ s, at which point the neck has disappeared.

The growth rate of the neck is a crucial observation in the fusion process. The initial neck growth depends on the balance between the capillary and viscous forces. For low viscosity liquids and short times, a logarithmic power law can describe the neck growth process, provided that $R_n \ll R_0$ [23]. This can be expressed as follows,

$$R_n(t) \propto t \ln t, \qquad (7.5)$$

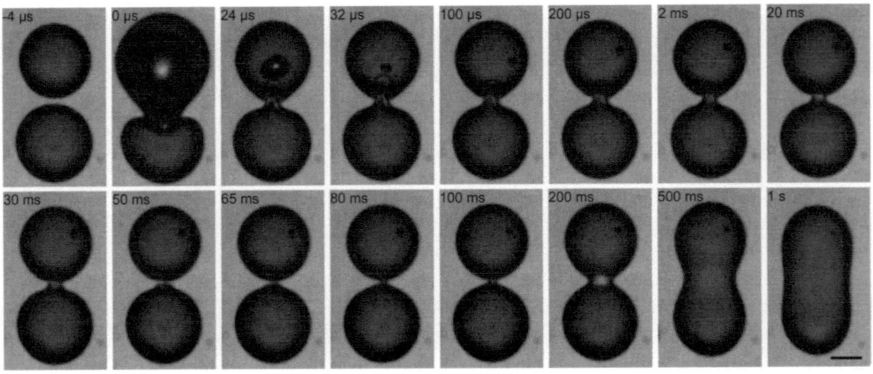

Fig. 7.4 Selected frames showing the fusion process of two nearly equal-sized droplets. The scale bar represents 40 μm. Reproduced from [21] with permission from the Royal Society of Chemistry

where R_n is the radius of the neck and R_0 is the radius of the fused droplet. Later, inertia becomes the dominant factor, and the growth rate follows a power law. However, when this model is compared to experimental data, it does not achieve satisfactory agreement due to the highly viscous continuous phase in this study. As a result, a modified model for neck growth based on purely viscous flow, i.e., the Stokes' equation, is proposed [24]. This model suggests that the entire evolution, except immediately after fusion, can be reasonably fitted to the modified equation, which is expressed as

$$R_n(t) = A + B\exp(-t/t_1), \tag{7.6}$$

where $A = -44.9$ μm, $B = -51.3$ μm, and $t_1 = 0.329$ s in this particular case. The neck growth rate decreases due to viscous dissipation. Figure 7.5 illustrates the evolution of the neck radius, along with the fitted result (Eq. (7.6)), depicted by the solid line. The reasonable agreement indicates that viscous dissipation, rather than inertial effects, predominates and slows down the neck growth. Therefore, the fusion process is influenced not only by the droplet size and the distance between the two droplets but also by the liquid properties, such as viscosity, density, and surface tension.

Furthermore, if the continuous phase between two droplets has not been completely drained, these droplets cannot fuse, as shown in Fig. 7.6. A cavitation bubble is created in the lower droplet, reaching its maximum at $t = 0$ μs. The bubble then starts to shrink and collapses at $t = 36$ μs. At $t = 20$ ms, the two droplets clearly detach, and by $t = 40$ ms, they assume spherical shapes due to interfacial tension. The primary reason for the lack of fusion is that the thin oil film between the two droplets does not drain completely.

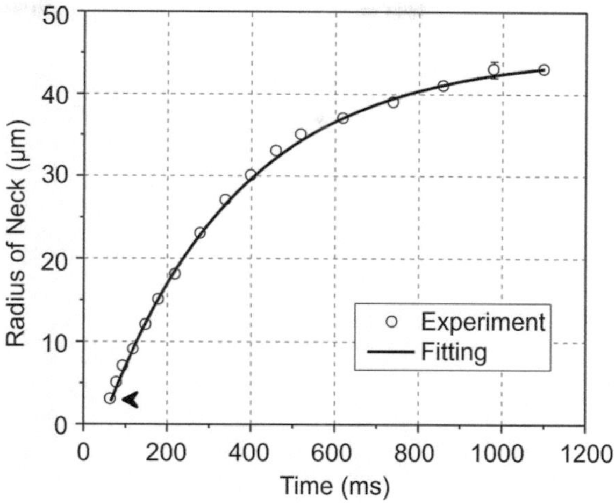

Fig. 7.5 Evolution of the neck radius after fusion occurs. Reproduced from [21] with permission from the Royal Society of Chemistry

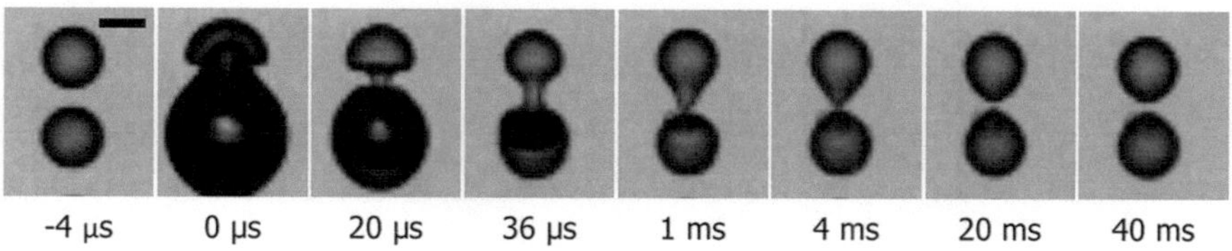

Fig. 7.6 Selected frames showing the un-fusion process of two nearly equal-sized droplets. The scale bar is 10 μm

7.4 Fusion of Two Unequal-Sized Droplets

The fusion process of two unequal-sized droplets is investigated by creating a cavitation bubble inside the smaller droplet, as shown in Fig. 7.7. The radii of the larger and smaller droplets are 18 μm and 8 μm, respectively. The oil film drainage and fusion initiation are similar to the case of two equal-sized droplets. However, when the neck between the droplets is established, the smaller droplet shrinks rapidly while the larger droplet grows. This behavior can be explained by the difference in interfacial pressure jump between the smaller and larger droplets. The pressure jump across the interface follows the Young–Laplace equation, which can be expressed as

$$P_{\text{droplet}} = P_{\text{oil}} + \frac{2\gamma}{R} \tag{7.7}$$

where P_{droplet} is the pressure inside the water droplet, P_{oil} is the pressure in the oil phase, γ is the surface tension between the water droplets and oil continuous phase, and R is the radius of the droplet. According to Eq. (7.7), the pressure within the smaller droplet is higher than that in the larger droplet. Consequently, when the

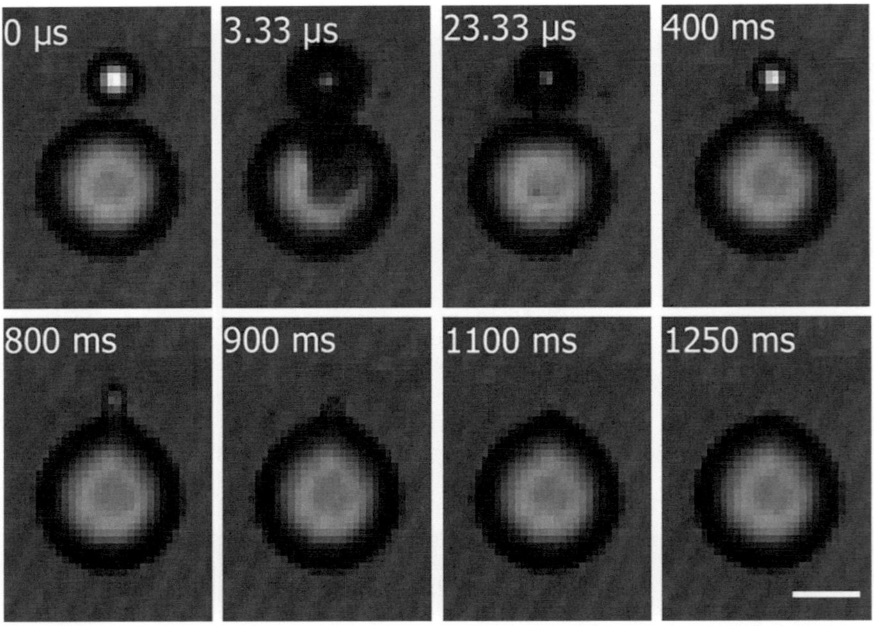

Fig. 7.7 Fusion process of two unequal-sized droplets. The scale bar is 20 µm. Reproduced from [21] with permission from the Royal Society of Chemistry

droplets fuse, a flow will be driven from the smaller to the larger droplet due to the pressure difference.

Unlike the fusion of two equal-sized droplets, the neck does not grow noticeably once it is formed. This observation enables the formulation of a simple model for the radial dynamics of the droplets, assuming the neck length and diameter remain constant. The flow from the smaller droplet (radius R_2) to the larger droplet (radius R_1) through the cylindrical neck is assumed to be laminar and fully developed. Therefore, the flow rate is given by

$$Q = \frac{\pi d^4 \gamma}{64 l \mu}\left(\frac{1}{R_2(t)} - \frac{1}{R_1(t)}\right), \tag{7.8}$$

where d and l represent the diameter and length of the neck, respectively, and μ is the viscosity of the liquid within the droplet. The flow rate balances with the rate of change of the droplet volumes,

$$Q = 4\pi R_1^2 \frac{dR_1}{dt} = -4\pi R_2^2 \frac{dR_2}{dt}. \tag{7.9}$$

From Eqs. (7.8) and (7.9), ordinary differential equations for the radii of both droplets are derived.

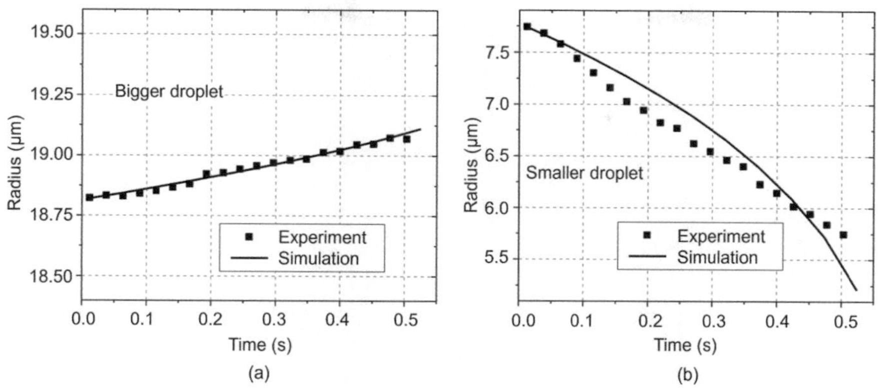

Fig. 7.8 Temporal evolution of the radii for **a** the larger droplet and **b** the smaller droplet. Reproduced from [21] with permission from the Royal Society of Chemistry

Figure 7.8 plots the measured radii of the droplets over time. The evolution is also predicted using Eqs. (7.8) and (7.9) with $l = 10$ μm, and $\mu = 2$ mPa s. The coefficient of interfacial tension $\gamma = 1.6 \times 10^{-4}$ N/m and the neck diameter $d = 2.85$ μm are varied to fit both curves. The overall model captures the growth and shrinkage of the two droplets, including the timescales. However, a more detailed model for the flow at the later stage of bubble shrinkage is necessary, especially as the droplet radius approaches the neck radius.

7.5 Fusion of Droplets in a Continuous Stream

The fusion of droplets transported in a flowing stream is also studied. The fusion dynamics of two equal-sized droplets moving in a continuous flow of droplets is presented in Fig. 7.9. The droplets travel at a velocity of 90 μm/s in the flow direction. A very similar fusion process dynamics is observed after the cavitation bubble is induced in the central droplet B, as shown in Fig. 7.9. Droplet B deforms the two neighboring droplets. However, despite the notable symmetry of the deformation, only the fusion of the central droplet B with the lower droplet C occurs. The upper droplet A continues to trail at a shorter distance. This is because the bubble is created slightly closer to droplet C within droplet B. It is unaffected by the flow direction or flow rate, as the bubble expansion velocity is several orders of magnitude greater than that of the droplet.

The strains of the central droplet B ($\Delta L/L$) in the flow (x-direction) and perpendicular (y-direction) directions are plotted in Fig. 7.10. The maximum strain in the y-direction occurs at $t = 4$ μs, measured as 3.2, while it is only 1.5 in the x-direction. When the bubble collapses at $t = 36$ μs, both strains equalize as droplet B returns to a circular shape. At later time ($t > 64$ μs), the strain in the y-direction becomes negative, as droplet B contracts in the flow direction. This unequal stretching of

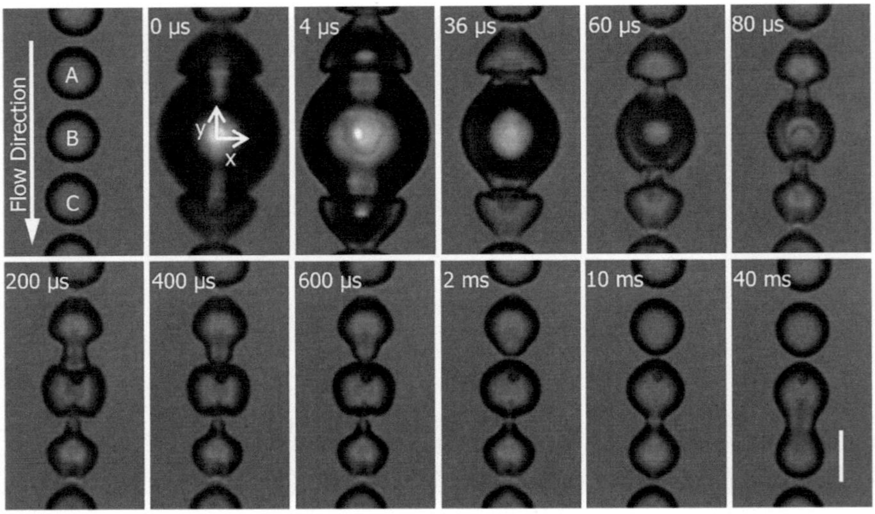

Fig. 7.9 Fusion process of two aqueous droplets in a flow stream inside the 100-μm channel. The scale bar is 40 μm. Reproduced from [21] with permission from the Royal Society of Chemistry

droplet B is due to the strong viscosity ratio (1250:1) between the dispersed and the continuous phase.

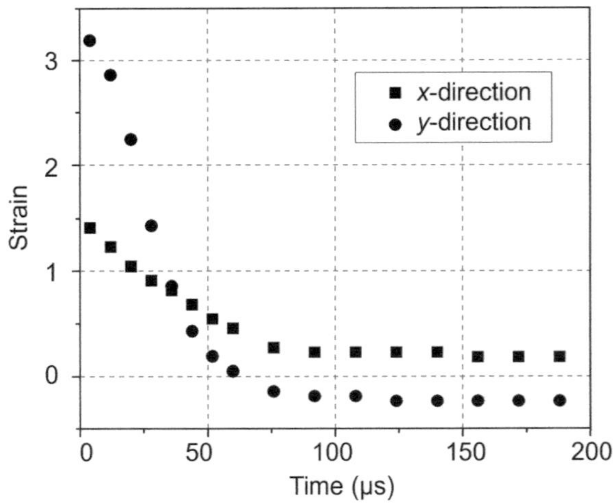

Fig. 7.10 Strain of the droplet B in x- and y-directions. Reproduced from [21] with permission from the Royal Society of Chemistry

7.6 Fusion of Droplets with Bubble/Cell

This section presents the bubble transfer between two droplets, the diffusion dominant mixing between the two fusing droplets and the fusion of droplets with an encapsulated cell.

Bubble transfer between two droplets

The break-up and transfer processes of the bubble between two equal-sized droplets are illustrated in Fig. 7.11. Two equal-sized droplets, A and B, are generated using a microfluidic channel with a T-junction. A cavitation bubble is created inside droplet B ($t = 0$ μs) using a focused laser pulse. Droplet B expands due to the expanding bubble, while droplet A is compressed and deformed by the expanding droplet B. The bubble reaches its maximum size, and the tip of the bubble penetrates the thin oil film between the two droplets at $t = 4$ μs. The bubble then splits into two parts at $t = 8$ μs. One part remains inside droplet B, while the other part is transported into droplet A from $t = 8$ μs to $t = 40$ μs. Unlike the fusion process described in the previous section, the tip of the bubble does not contract back into droplet B but instead breaks apart. Half of the bubble remains in droplet B, while the other half is trapped in droplet A. In this process, materials from droplet B can be successfully transferred to droplet A using the broken bubble, with the two droplets remaining separate after the transfer. This demonstrates great potential for applications such as mass transfer between droplets and pumps driven by pulsed laser-induced cavitation bubbles.

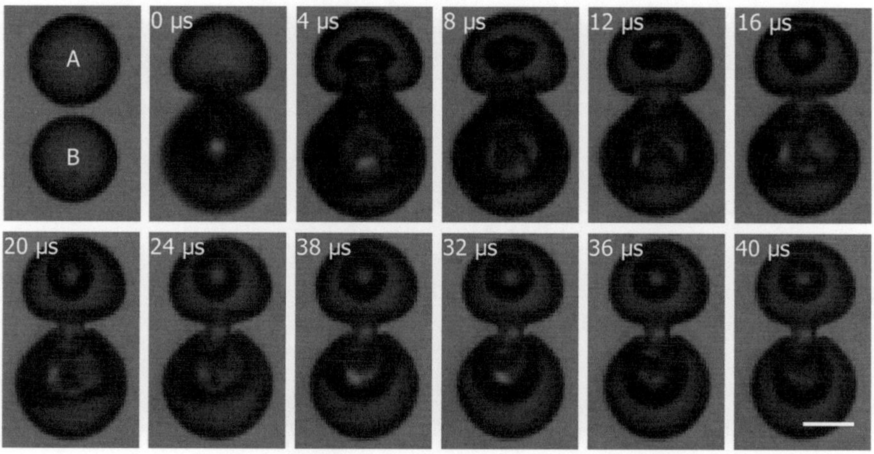

Fig. 7.11 Selected frames depicting the bubble transfer process between two equal-sized droplets. The scale bar is 40 μm

Diffusion mixing of two droplets

The diffusion-dominated mixing process between two fusing droplets is illustrated in Fig. 7.12. The sequence of the mixing process is shown with two different-colored droplets: one made of aqueous red dye and the other containing DI water. A modified microfluidic chip for droplet generation is used, featuring two T-junctions combined into the same microchannel to generate two different-colored droplets simultaneously. This setup produces a continuous stream of equal-sized droplets of alternating colors for larger droplet volumes. These droplets span the entire channel and become significantly deformed into Taylor droplets [25]. Subsequently, when a cavitation bubble is created just above the thin oil layer, the expanding bubble (not shown in the figure) injects a small portion of the red-dyed fluid from the upper droplet into the lower droplet at $t = 100\ \mu s$. The oil layer ruptures, and the dye gradually diffuses from the upper droplet into the lower droplet, and vice versa. No residual liquid flow is observed shortly after the bubble has vanished. Thus, diffusion can be described using Fick's law, which is expressed as

$$\frac{\partial c}{\partial t} = D\nabla^2 c, \tag{7.10}$$

where c is the mass concentration, t is the time and D is the constant of mass diffusivity. If a stepwise change in the one-dimensional concentration field at $x = 0$ is

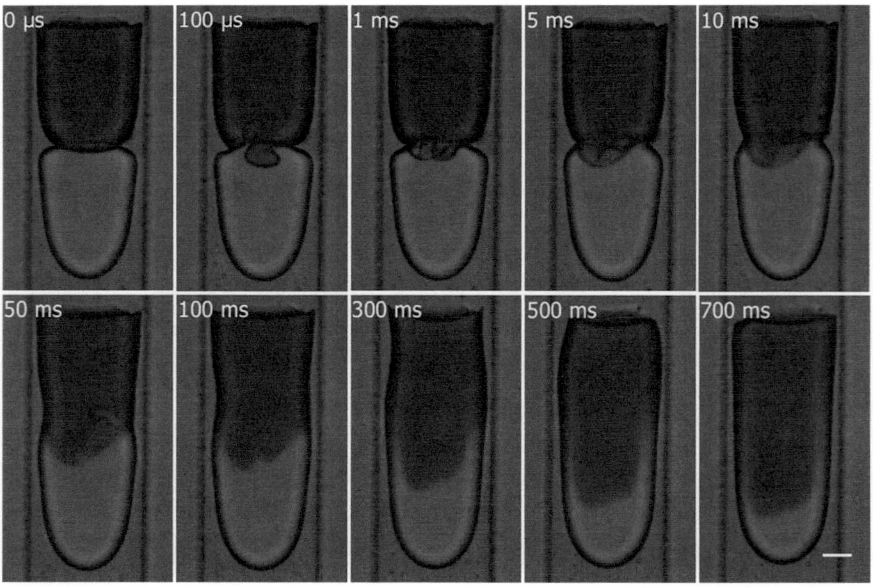

Fig. 7.12 Mixing process of two aqueous droplets: the upper droplet contains aqueous red dye, while the lower droplet contains DI water. The scale bar is 40 μm. Reproduced from [21] with permission from the Royal Society of Chemistry

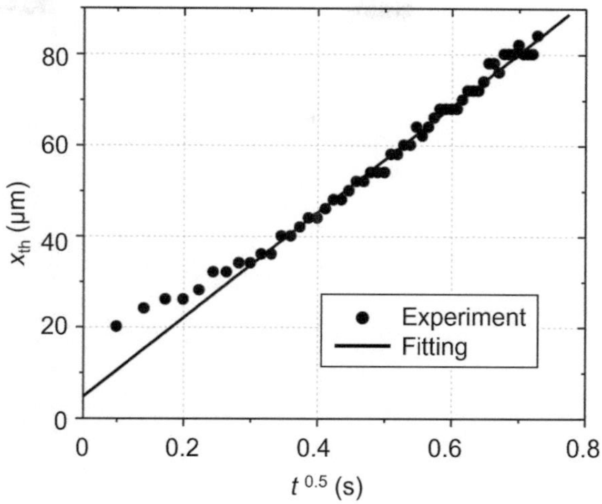

Fig. 7.13 Position of x_{th} as a function of the square root time

imposed, the diffusion process (in infinite space) follows

$$x_{th} \propto \sqrt{t}, \tag{7.11}$$

where x_{th} denotes the position of the threshold concentration. In Fig. 7.13, the evolution of the measured concentration field is fitted to a square root of t. During the measurement period, there is reasonable agreement except for the initial period. This discrepancy is due to the remaining fluid motion ($t = 100$ μs in Fig. 7.12) resulting from the film rupture between the two droplets, which may lead to the observed enhanced mixing. Nevertheless, the agreement indicates that the mixing within the fused droplets is mainly governed by diffusion and can thus be described by Fick's law. It is also important to note that, to accurately predict the diffusion process at later times, boundary effects must be considered.

Fusion of droplets with single cells

In cell analysis, it is critical to control the volume of the reagents. The proposed fusion method allows for introducing quantified reagents into droplet reactors which encapsulate a single cell. The on-demand fusion technique is demonstrated in Fig. 7.14. The micrographs selected from a high-speed recording present the successful fusion of one droplet with a single HepG2 (Hepatocellular carcinoma, human) cell with the second droplet containing the magenta ink dye. The minimum distance between the two droplets is measured as 4 μm. The white dash circles follow the movement of the HepG2 cell and the black dash circle shows the location of the laser-induced cavitation bubble. Both droplets are stagnant in the droplet collection chamber. The bright spot at the laser focus is due to the continuum emission from the short-lived plasma created by the focused laser.

The laser pulse is focused into the droplet without the cell at $t = 0$ μs. The white point is the hot plasma created by the focused laser pulse and then one cavitation

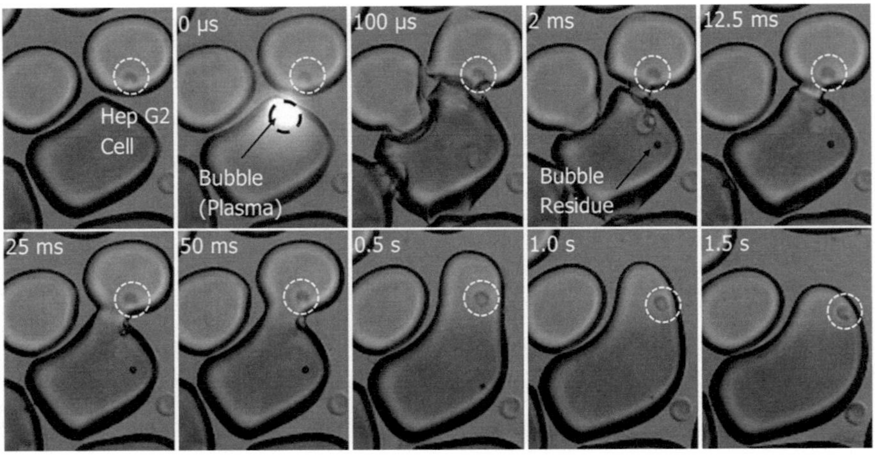

Fig. 7.14 Fusion process of two droplets containing HepG2 cell and aqueous dye. Reproduced from [21] with permission from the Royal Society of Chemistry

bubble is formed. The continuous phase between the two droplets is again drained by the cavitation bubble at $t = 100$ μs. Then, the connection neck between the two droplets is formed at $t = 2$ ms. Under the mild driving force from interfacial tension, both droplets fuse and mix gradually from $t = 25$ ms to $t = 1.5$ s.

During the fusion process, the permanent poration of the cell membrane may lead to cell death when the cell is in vicinity to the expanding bubble. However, in this case, it is not expected that the destructive effect of bubble dynamics on the cell. This is because the oil film between the droplets dampens the bubble dynamics, and the cell is located in the distant droplet. Cell apoptosis and proliferation could be investigated in future.

The laser-induced cavitation bubble also has great potential in cell fusion. The fluidic jet generated by the collapse of the bubble in vicinity to a solid boundary can be used to penetrate a cell's membrane to trigger fusion process. However, more experiments are necessary to find the optimized parameters for the poration of the cell membrane and the fusion processes.

7.7 Summary

Fast on-demand droplet fusion is achieved via the transient and rapid flow from a short-lived laser-induced cavitation bubble within one of the droplets. The theoretical analysis investigates the droplet fusion mechanism using laser-induced cavitation bubbles and the neck growth theory between two fusing droplets. Unlike other microfluidic fusion techniques, this developed method can be used for the fusion of both equal-sized and unequal-sized droplets, as well as for droplet fusion in a continuous stream. Potential applications include the mixing of two reagents and the merging of droplets containing encapsulated cells. This droplet fusion technique offers several advantages, such as good controllability in time and space, simple fabrication requiring only optical access, and extremely fast actuation enabling rapid droplet fusion.

References

1. Moragues T, Arguijo D, Beneyton T, Modavi C, Simutis K, Abate AR, Baret J-C, deMello AJ, Densmore D, Griffiths AD (2023) Droplet-based microfluidics. Nat Rev Methods Primers 3:32
2. Ding Y, Howes PD, deMello AJ (2020) Recent advances in droplet microfluidics. Anal Chem 92:132–149
3. Joanicot M, Ajdari A (2005) Droplet control for microfluidics. Science 309:887–888
4. Suea-Ngam A, Howes PD, Srisa-Art M, deMello AJ (2019) Droplet microfluidics: from proof-of-concept to real-world utility? Chem Commun 55:9895–9903
5. Tan YC, Fisher JS, Lee AI, Cristini V, Lee AP (2004) Design of channel geometries for the control of droplet volume, chemical microfluidic concentration, and sorting. Lab Chip 4:292–298
6. Tan YC, Ho YL, Lee AP (2007) Droplet coalescence by geometrically mediated flow in microfluidic channels. Microfluid Nanofluid 3:495–499
7. Liu K, Ding HJ, Chen Y, Zhao XZ (2007) Droplet-based synthetic method using microflow focusing and droplet fusion. Microfluid Nanofluid 3:239–243
8. Hung LH, Choi KM, Tseng WY, Tan YC, Shea KJ, Lee AP (2006) Alternating droplet generation and controlled dynamic droplet fusion in microfluidic device for CdS nanoparticle synthesis. Lab Chip 6:174–178
9. Köhler JM, Henkel T, Grodrian A, Kirner T, Roth M, Martin K, Metze J (2004) Digital reaction technology by micro segmented flow—components, concepts and applications. Chem Eng J 101:201–216
10. Fidalgo LM, Abell C, Huck WTS (2007) Surface-induced droplet fusion in microfluidic devices. Lab Chip 7:984–986
11. Li L, Boedicker JQ, Ismagilov RF (2007) Using a multijunction microfluidic device to inject substrate into an array of preformed plugs without cross-contamination: comparing theory and experiments. Anal Chem 79:2756–2761
12. Chen DL, Li L, Reyes S, Adamson DN, Ismagilov RF (2007) Using three-phase flow of immiscible liquids to prevent coalescence of droplets in microfluidic channels: criteria to identify the third liquid and validation with protein crystallization. Langmuir 23:2255–2260
13. Ahn K, Agresti J, Chong H, Marguez M, Weitz DA (2006) Electrocoalescence of drops synchronized by size-dependent flow in microfluidic channels. Appl Phys Lett 88:264105
14. Priest C, Herminghaus S, Seemann R (2006) Controlled electrocoalescence in microfluidics: targeting a single lamella. Appl Phys Lett 89:134101
15. Wang J, Lu C (2006) Microfluidic cell fusion under continuous direct current voltage. Appl Phys Lett 89:234102
16. Singh P, Aubry N (2007) Transport and deformation of droplets in a microdevice using dielectrophoresis. Electrophoresis 28:644–657
17. Schwartz JA, Vykoukal JV, Gascoyne PRC (2004) Droplet-based chemistry on a programmable micro-chip. Lab Chip 4:11–17
18. Tan WH, Takeuchi S (2006) Timing controllable electrofusion device for aqueous droplet-based microreactors. Lab Chip 6:757–763
19. Tresset G, Takeuchi S (2005) Utilization of cell-sized lipid containers for nanostructure and macromolecule handling in microfabricated devices. Anal Chem 77:2795–2801

20. Lorenz RM, Edgar JS, Jeffries GDM, Chiu DT (2006) Microfluidic and optical systems for the on-demand generation and manipulation of single femtoliter-volume aqueous droplets. Anal Chem 78:6433–6439
21. Li ZG, Ando K, Yu JQ, Liu AQ, Zhang JB, Ohl CD (2011) Fast on-demand droplet fusion using transient cavitation bubbles. Lab Chip 11:1879–1885
22. Quinto-Su PA, Kuss C, Preiserb PR, Ohl CD (2011) Red blood cell rheology using single controlled laser-induced cavitation bubbles. Lab Chip 11:672–678
23. Hartland S, Yang B, Jeelani SAK (1994) Dimple formation in the thin film beneath a drop or bubble approaching a plane surface. Chem Eng Sci 49:1313–1322
24. Hiram Y, Nir A (1983) A simulation of surface tension driven coalescence. J Colloid Interface Sci 95:462–470
25. Mandal TK, Das G, Das PK (2008) An experimental study of Taylor bubbles and Taylor droplets in liquid-liquid system. Ind Eng Chem Res 47:7048–7057

Chapter 8
Optofluidic Measurement with Imaging Techniques

8.1 Introduction

Over the past decades, various miniaturized optical components fabricated on a chip have been extensively explored across different platforms for biochemical detection and analysis. A key advantage in optofluidics is the exceptional size compatibility between biological objects (e.g., bacteria, cells, or organelles), microfluidic channels, and optical beams in the visible or near-infrared spectrum, this alignment enables precise tailoring of microchannel dimensions and optical beam diameters to achieve desired fluidic and optical effects. Light carries information encoded with its basic properties, such as phase, electromagnetic frequency, intensity, or polarization. The interaction of light with the inspected sample can modulate these properties, which need to be decoded by specific tools such as interferometers, spectrometers, photometers, and polarization analyzers. Meanwhile, these analytical instruments are successfully integrated with novel photonic capabilities in a miniaturized system. Optofluidic technologies are revolutionizing bioanalysis by pushing performance boundaries, enhancing sensitivity and selectivity, and fulfilling critical requirements for point-of-care devices. Numerous efforts have been invested in label-free optical biosensors, covering various types of optical measurements such as transmission, emission, adsorption, polarimetry, and light scattering. These sensing techniques show potential applications not only in medical diagnostics but also in drug discovery, food safety, and environmental monitoring.

Imaging is likely the most fundamental and widespread microscopic technology. It captures the topographic information of observed objects under illumination. This technique is simple, easy to use, cost-effective, and allows real-time observations. When coupled with high-speed camera imaging, it enables researchers to obtain time-resolved information down to the sub-millisecond level. Additionally, it is one of the easiest techniques to integrate with an optofluidic device, as most materials used in their fabrication are transparent, facilitating straightforward optical imaging.

S. Xiong and A.-Q. Liu, *Optofluidics: Fundamentals and Applications*,
Microfluidics and Nanophotonics: Science and Engineering 4,
https://doi.org/10.1007/978-981-95-6296-1_8

In the following sections, we will introduce various measurement methods utilizing optofluidic imaging.

8.2 Refractive Index Measurement

A cell is the fundamental functional unit of living organisms and the smallest self-sustaining biological entity. Since its discovery by Robert Hooke in 1665, scientists have extensively researched living cells. Recent advancements in optics and photonics provide biophysicists with new methods to study cells. Microscopic systems such as electron microscopy and fluorescence microscopy have been employed to examine subcellular structures, while optical tweezers are used for cell manipulation. Lasers have been developed for biomedical applications with low phototoxicity on cells. Similarly, photodetectors and spectrometers are used to obtain optical information about cells based on several optical quantities, such as scattering, absorption, and refractive index. These quantities are associated with cell shape, intracellular content, and size. Optical techniques offer the advantage of non-contact measurement methods, avoiding interference with the sample cell. In this book, the primary measurement focus is the cell refractive index, which plays a vital role in biophysics and biomedical fields.

Refractive index is an optical parameter used to quantify how fast a light wave travels through a medium. The higher the refractive index of a material, the slower the light travels within it. This parameter is often used to identify substances and measure their purity or concentration. For example, the refractive index can be used to determine the concentration of sugar in a solution, which is associated with the Brix level. The refractive index of a solution, n, is linearly related to the concentration of the solute, C (in g per 100 ml), as

$$n = n_0 + \alpha C, \tag{8.1}$$

where n_0 is the refractive index of the solvent and α is a constant known as the specific refraction increment that depends on the nature of the solute [1].

Interpreting the refractive index value for a non-uniform medium, such as a living cell, is more complex than for a uniform medium like a solution. Cells contain various biomolecular components, including inorganic salts, nucleic acids, proteins, carbohydrates, lipids, and water. However, for most cells, the primary intracellular constituents are proteins, and cells can be regarded as capsules filled with protein solution [2]. According to Eq. (8.1), the refractive index of a solution depends on the concentration of the solutes and the specific refraction increment. The relationship between the concentration of proteins in a cell and the cell's refractive index can only be established using Eq. (8.1) if different proteins have the same specific refraction increment. The specific refraction increment, α, for many proteins has been measured because it is required for determining molecular weight based on light scattering measurements and for calibrating interferometer systems in ultracentrifuges

Table 8.1 Specific refraction increments of intracellular constituents [3]

Constituents	Specific refraction increment
Human serum albumin	0.00185
Human gamma globulin	0.00188
Human serum globulin	0.00186
Human α_2 globulin	0.00183
Human β_1 globulin	0.00185
Human euglobulin	0.00183
Human pseudoglobulin	0.00181
Lactoglobulin	0.00182
β lactoglobulin	0.00182
Peptone	0.00183
Glycine	0.00179

or electrophoresis systems. The specific refraction increments of some intracellular constituents are listed in Table 8.1.

The table shows that the average specific refraction increment for soluble proteins is 0.00185 $\pm$ 2%. Similarly, nucleic acids, nucleoproteins, and protein breakdown products like peptones and amino acids have values close to those of proteins. Given that these other constituents are present in much smaller amounts compared to proteins, the specific refraction increment for cells can be regarded as 0.00185. By substituting this value into Eq. (8.1) and considering n_0 as the refractive index of water or a dilute salt solution depending on the extracellular medium, the total concentration of organic solids in the cell can be determined when the cell's refractive index is known.

Using the concentration of the organic solids, the mass density of water (in g per 100 ml) can be calculated as

$$\rho_{\text{water}} = 100 - \nu C, \tag{8.2}$$

where ν is the specific volume (the volume occupied by 1 g of the solute in the solution) of proteins, with a value of 0.75 [1]. Using these two parameters, the mass of the cell can be determined if the cell volume is known. The cell volume can be calculated as

$$V = (4\pi / 3) r_c^3, \tag{8.3}$$

where r_c is the cell radius, with the assumption that the cell is spherical. The dry mass (organic solids) and water mass within the cell can be calculated as

$$M_{\text{dry}} = VC, \tag{8.4a}$$

and

$$M_{\text{water}} = V\rho_{\text{water}}. \tag{8.4b}$$

Subsequently, the mass of cell is the summation of the dry mass and water mass of the cell [4].

Applications in cell biology and disease diagnoses

By monitoring the cell's refractive index, changes in the solid content within the cell can be continuously quantified throughout its cycle. This technique acts as an indicator to study and control cell states during the division cycle and assess the effects of extracellular conditions in a controlled environment. It provides a non-invasive, real-time measurement method for long-term cell monitoring.

Living cells are highly sensitive to changes in the extracellular environment, especially osmotic pressure. Deviations from optimal osmotic pressure can cause cells to shrink or swell. Cell biologists seek to elucidate the relationship between cell volume and osmotic pressure, as well as cell viability under varying osmotic conditions. Refractometers are indispensable for analyzing osmotic responses [5]. In hematology, refractometry plays a vital role in determining hemoglobin concentration in red blood cells. Since hemoglobin is the primary component of mammalian red blood cells, refractometry effectively correlates the refractive index of a red blood cell with its hemoglobin concentration [6]. Refractometry also holds significant value in oncology [7]. Normal cells typically exhibit an refractive index between 1.35 and 1.37, whereas certain cancer cells, such as HeLa cells, display a slightly higher refractive index of 1.392. This distinction can be pivotal for early cancer diagnosis. Malignant cancer cells often have larger nuclei and higher solid concentrations, resulting in an elevated refractive index.

Techniques for refractive index measurements

Researchers have developed various optical systems to measure the cell refractive index, such as optical fiber and microscopic systems. Optical fiber systems measure single living cells using optical signals like phase measurement, light scattering intensity, localized surface plasmon resonance, or resonance condition variation in an optical cavity. In microscopic systems, the phase shift caused by the cell or tissue on the transmitted wavefront is measured, and the cell refractive index is decoupled using image algorithms. This section explores various techniques for measuring refractive index.

Surface plasmon techniques

Surface plasmon resonance on metallic surfaces is the most renowned and extensively utilized label-free optical bioanalytical technique for accurately measuring cell refractive index [8]. Surface plasmons are electromagnetic waves that travel parallel to the metallic-dielectric interface. These oscillations are highly sensitive to the environment at the boundary between the metal and the external medium, making them ideal for biosensing applications. The excitation of surface plasmons on planar metallic surfaces is referred to as surface plasmon resonance, whereas on nano-sized metallic structures it is known as localized surface plasmon resonance.

For example, a virtual near-field probe, formed by the optical excitation of localized surface plasmons using radial polarization beams, is employed to measure the refractive index map of cell-substrate contacts [9]. The substrate comprises a glass cover slip coated with 50 nm of gold, to which the cell sample is attached. A total internal reflection fluorescence microscope system focuses the laser beam for precise surface plasmon localization. The reflected beam, captured by the CCD camera, reveals annular rings whose normalized radius converts into the corresponding surface plasmon resonance angle, thereby obtaining the local effective refractive index. A 2D refractive index map is produced by scanning the virtual probe across the sample area.

In another work, a nano-optical probe for real-time cell refractive index measurement was developed using a tapered fiber tip coated with 13-nm gold nanoparticles [10]. The evanescent wave at the tip excites localized surface plasmon resonance. When the tip is dipped into solutions of varying refractive indices, the reflected intensity changes accordingly, forming the basis for measurement. Its sensitivity is 5400% intensity change per RIU, with a resolution of 3.7×10^{-5}. However, the small tip size (80 nm) necessitates a complex scanning mechanism on the microscope to map a single cell's refractive index, and the tip must be dipped into a reference medium before measuring real samples, complicating the procedure with a movable probe on an attached sample.

Light scattering techniques

Light scattering intensity serves as an alternative optical parameter to measure the refractive index of a sample. This phenomenon occurs when light rays are deflected in random directions due to irregularities within the medium or at the boundary between two media. Analyzing the scattering pattern of light at a specific wavelength on a sample medium, based on the scattering angle, provides valuable structural, morphological, or optical insights about the sample. In the case of studying living cells, which are generally spherical, the Mie scattering solution is applied as an analytical method for solving Maxwell's equations to determine the scattering pattern of these spherical entities.

To determine the size and refractive index of a spherical sample, the scattering data must be empirically inverted by creating a model and adjusting its parameters to achieve the best fit [11]. An alternative empirical method is the optical strip-map technique [12], which provides size and refractive index measurements under 1 ms. This method has been used to characterize bacterial spores and *Escherichia coli* with a precision of 0.01 RIU and 0.6% in size. However, both methods require measuring scattering intensities at multiple angles, complicating the measurement system with the need for several light sources and detectors.

To eliminate the need for scattering data at multiple angles, a two-beam optical trap based on forward scattering intensity measurement has been developed to determine cell size and refractive index [13]. This system uses a counter-propagating optical trap to hold a single cell within a microfluidic channel. The cell's refractive index and size are measured through the forward scattered light at different positions by adjusting the power ratio between the two laser beams. The system offers a resolution of 0.013 RIU and 3.3% for size measurement, which is comparatively lower than that of microscopic techniques.

Optical resonant techniques

Optical resonators are a class of highly sensitive optical devices used as biosensors for biomolecules or single living cell measurements. They come in two main types: standing wave resonators, such as Fabry-Pérot (FP) resonator, and traveling wave resonators, such as ring resonators. The FP resonator consists of reflecting surfaces that create a standing wave cavity for light waves, while the ring resonator features a closed-loop waveguide where traveling light waves are coupled, intensifying through constructive interference over multiple round trips. The resonant conditions of these light waves are influenced by the refractive index of the medium within the FP cavity or surrounding the ring resonator, making these optical properties useful for sensing applications.

In sensing applications, optical resonators are categorized into two types: surface sensors and volume sensors. Surface sensors operate on the Whispering Gallery resonant mode of ring or spherical cavities [14, 15]. Biomolecules adhere to the surface and interact with the evanescent wave, altering the resonant mode for measurement. The interaction depth is typically less than 1 μm, ideal for protein or DNA mass measurements. Conversely, for single-cell measurements that require a deeper interaction depth, volume sensors are more suitable. Volume sensors detect changes in light propagation paths within micro-optical cavities [16, 17]. When cells are introduced into a microcavity (e.g., a laser or Fabry-Pérot cavity), they modify the resonant mode by changing the optical path. A microfluidic system for measuring cell refractive index using a laser cavity has been developed [7]. Single cells flow into the analysis region where a laser beam scans the cell stream. The cell's presence alters the optical path length of the laser cavity, causing frequency shifts and intensity modulation based on semiconductor laser theory. By monitoring these optical signals, the cell refractive index is determined. This system measures various cell types with a precision of 0.004 RIU. A Fabry-Pérot resonator-based microfluidic system has been developed for cell refractive index measurement [18]. A single cell is physically captured and fixed in the resonant cavity, where the differential buffer method records spectral shifts with two different buffers. These spectral shifts determine the cell's refractive index and size. This system, demonstrated to measure Madin Darby canine kidney cells with a precision of 0.003 RIU, can be enhanced by improving the resonant cavity's finesse.

Microscopic techniques

Digital holographic microscopy is a method that offers phase measurements of living cells with interferometric resolution [19, 20]. The measured phase shift provides information about the cell's refractive index and thickness. Through quantitative phase images, both the integral refractive index and cellular thickness can be independently measured using the decoupling protocol. This protocol involves obtaining two holograms of the living cell with buffers of different refractive indices but the same osmolarity. Digital holographic microscopy boasts a low acquisition time of up to 20 μs and allows measurement of multiple cells in a single hologram. However, its resolution is limited to 0.01 in refractive index measurement and 1 μm in size measurement.

Hilbert phase microscopy, similar to digital holographic microscopy but with different phase retrieval processes [21], encodes the phase distribution of a microscope image using a plane-wave reference field for spatial modulation. A quantitative phase image is then acquired via a spatial Hilbert transform. For cell measurements, living cells are confined in a microchannel. The cell refractive index is obtained by performing a relative phase measurement on both the culture medium ($n = 1.337$) and the cell. This technique achieves a precision of 0.005 RIU with a low acquisition time of 3 ms. However, it requires cell confinement, which might affect the cells, and does not measure cell size.

Confocal microscopy is frequently integrated with other techniques to improve the accuracy of cell refractive index measurements [22]. In immersion refractometry, the tissue sample and immersion liquid are positioned between glass slides, ensuring uniform thickness. The confocal microscope measures thickness with a resolution of 0.625 μm. The refractive index contrast generates an axial scaling factor, derived from the known thickness, enabling a measurement precision of 0.004. Another method involves combining confocal microscopy with quantitative phase-amplitude microscopy [23], similar to digital holography, to obtain a phase image and decouple the tissue refractive index from its thickness. While confocal microscopy enhances precision, it requires sample confinement and is costly.

Optofluidic measurements

Advances in lab-on-a-chip technology now enable the development of integrated optical systems capable of measuring the refractive index of single cells. These systems also incorporate additional functions, including buffer injection and exchange, cell sorting, encapsulation, and trapping. Here, an example is provided for measuring cell refractive index with an on-chip optofluidic imaging system.

The optofluidic imaging system integrates a microfluidic chip for cell manipulation and liquid medium modulation with a phase-contrast microscope to perform immersion refractometry. Aqueous samples containing cells are introduced into the microchannel, where an array of trapping sites immobilizes individual cells within a designated sample area. By immersing the trapped cells in a controlled liquid medium, the system measures their refractive index through optical contrast analysis, as illustrated in Fig. 8.1.

When light traverses a single cell within the microfluidic chip immersed in a medium of lower refractive index, the rays passing through the cell experience a phase delay relative to the background light passing through the surrounding medium. The optical path difference (OPD) is quantified by $\mathrm{OPD} = (n_{\mathrm{cel}} - n_{\mathrm{med}})\cdot t_{\mathrm{cel}}$, where t_{cel} is the cell thickness, n_{cel} and n_{med} are the refractive indices of the cell and medium, respectively. However, bright-field microscopy exhibits limited sensitivity to this phase delay. To enhance refractive index measurement sensitivity, phase-contrast microscopy is employed. This technique introduces a quarter-wavelength ($\lambda/4$) phase shift, maximizing intensity contrast between the cell and medium. Depending on the

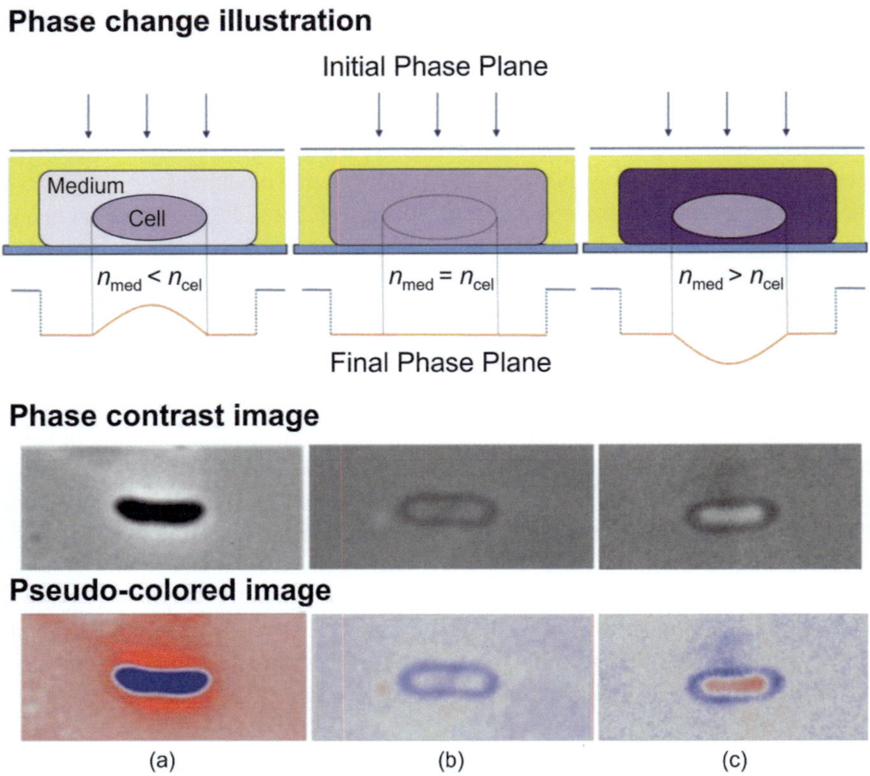

Fig. 8.1 Working principle of the null method in immersion refractometry. The images demonstrate three scenarios: **a** medium refractive index (n_{med}) < cell refractive index (n_{cel}), **b** $n_{med} = n_{cel}$, and **c** $n_{med} > n_{cel}$. Reproduced from [24] with permission from the Royal Society of Chemistry

relative refractive index between the cell and the surrounding medium, phase-contrast microscopy reveals distinct visual characteristics. When the external medium has a refractive index lower than that of the cell ($n_{med} < n_{cel}$), the cell appears darker as shown in Fig. 8.1a. Whereas when the external medium has a higher refractive index ($n_{med} > n_{cel}$), the cell appears brighter as shown in Fig. 8.1c. Crucially, when their indices match ($n_{med} = n_{cel}$), the cell appears to be invisible as shown in Fig. 8.1b, thereby enabling refractive index measurement through the null condition. To implement this null method, an integrated chaotic micromixer dynamically tunes the refractive index of medium from low to high by adjusting the flow ratio of two solutions. For example, DI water ($n = 1.3326$) and Ficoll solution ($n = 1.4651$) enable continuous refractive index adjustment of the external medium between 1.333 and 1.465 for bacterial cell measurements [24].

System sensitivity is governed by two factors: the phase-shift detection limit of the phase-contrast microscopy and the resolution of the commercial refractometer used to measure the refractive index of the medium. For a non-absorbing phase plate, phase changes below λ/100 are readily detectable. Given a typical bacterial thickness of 1 μm and detection wavelength of 514 nm, this translates to a minimum refractive index difference (Δn) of ~0.005. The detection limit of the phase-contrast microscopy depends on the light absorption of the phase plate, in which $\lambda/1000$ can be achieved with heavily absorbing plates. The sensitivity can be theoretically improved to 0.0005 [3]. Since the refractometer achieves a 10^{-4} resolution in medium measurements, its precision does not constrain overall system sensitivity.

Trapped cells within the optofluidic chip undergo phase-contrast imaging via a digital CCD camera for biophysical characterization. Size and shape parameters (length, width, aspect ratio) are quantified by analyzing captured images using standard software (e.g., ImageJ). For refractive index measurement, sequential images are acquired under varying external media conditions, followed by intensity profiling to quantify contrast differences between intracellular regions and the surrounding medium.

This method was used to investigate three bacteria species: *Escherichia coli*, *Shigella flexneri* and *Vibrio cholerae*. A sample size of 250 cells per species is measured and statistically characterized as shown in Fig. 8.2. Both *E. coli* and *S. flexneri* share a similar morphology, being rod-shaped bacteria. According to these results, *E. coli* exhibits mean dimensions of 2.83 μm (length) × 0.86 μm (width) with an aspect ratio of 3.87, while *S. flexneri* measures slightly smaller at 2.74 μm × 0.77 μm and an aspect ratio of 3.48. In contrast, *V. cholerae* is smaller and has a shape of comma, with a mean length of 1.21 μm, width of 0.43 μm, and aspect ratio of 2.84.

The optofluidic immersion refractometer is characterized using a PDMS testing structure. Phase-contrast images are captured while varying the external medium's refractive index from 1.410 to 1.414. Intensity contrast for each image is quantified as the ratio of the average pixel intensity of the PDMS testing structure to that of the surrounding medium (Fig. 8.3). When the external medium has lower refractive index as compared to the one of PDMS, the PDMS testing structure appears to be brighter and protruding. On the other hand, when the external medium has higher refractive index as compared to the one of the PDMS, the PDMS testing structure appears to be darker and concave. A phase inversion occurs between 1.410 and 1.414. Based on the interpolation of the measured data, the PDMS testing structure has a refractive index of 1.412. This value is experimentally confirmed by injecting a medium with $n = 1.412$, where the PDMS structure becomes optically invisible (inset of Fig. 8.3).

Figure 8.4 shows the pixel intensity analysis for single bacterium by varying the refractive index of the external medium. The intensity contrast is lower than 1 when the refractive index of the bacterium exceeds that of the external medium, and the relationship reverses otherwise. In Fig. 8.4a, the matching refractive index of 1.388 occurs when the intensity contrast is equal to 1, which is also the effective refractive index of the *E. coli*. The refractive index value aligns with light scattering measurements ($n = 1.390$) [25]. Parallel measurements revealed an effective refractive index of 1.422 for *S. flexneri* (Fig. 8.4b) and 1.365 for *V. cholerae*, demonstrating the technique's capacity to differentiate species based on biophysical properties.

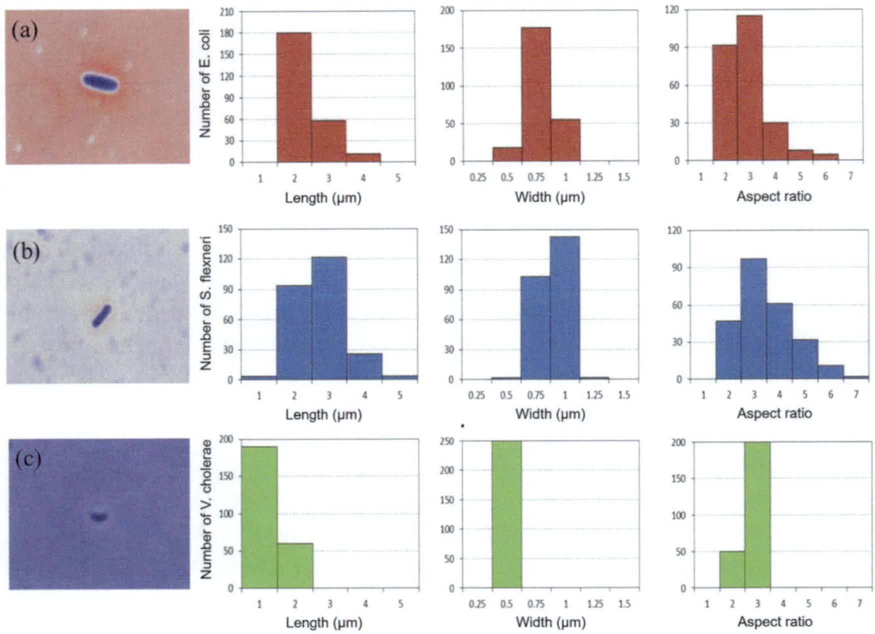

Fig. 8.2 Morphological measurements of **a** *Escherichia coli*, **b** *Shigella flexneri*, and **c** *Vibrio cholerae*. Reproduced from [24] with permission from the Royal Society of Chemistry

Statistical analysis of morphological parameters (size, shape) and effective refractive indices demonstrates distinct biophysical signatures across the bacterial species (e.g., *E. coli*, *S. flexneri*, *V. cholerae*). Species differentiation is achievable at a resolution of 5×10^{-3} RIU through optofluidic refractometry. When combined with dimensional metrics (e.g., *E. coli*: 2.83 μm length, 0.86 μm width, aspect ratio 3.87), the biophysical signatures establish a label-free identification framework for waterborne bacteria in treated water systems. The technique exhibits particular utility in rapid detection of *E. coli*, which is an indicator of fecal pollution and potential enteric pathogens presence. By enabling real-time quarantine of polluted water sources, this method mitigates outbreak risks while bypassing traditional culture-based delays.

8.3 Tensile Strength Measurement

The mechanical characterization of soft materials including water, biological tissues, cells, and gels has emerged as a critical research focus in contemporary engineering and biomedical fields, with applications spanning from artificial tissue engineering to flexible electronics. Among these properties, tensile strength represents a fundamental mechanical parameter. Water is known to exist in a metastable state when

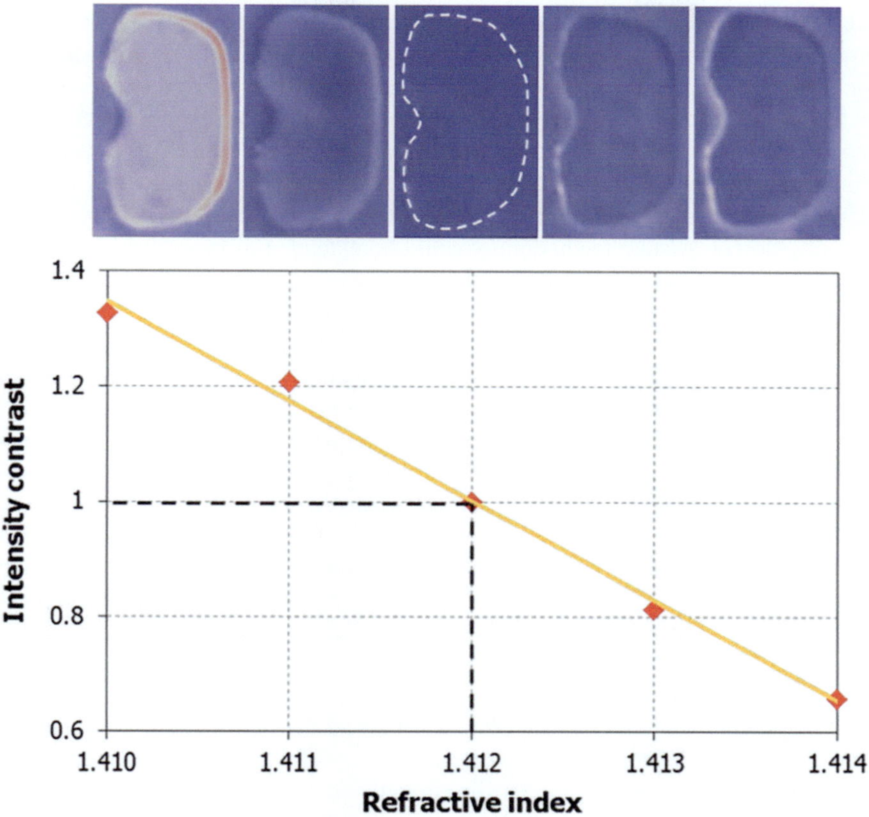

Fig. 8.3 Characterization of the optofluidic immersion refractometer by measuring the refractive index of PDMS testing structure. Reproduced from [24] with permission from the Royal Society of Chemistry

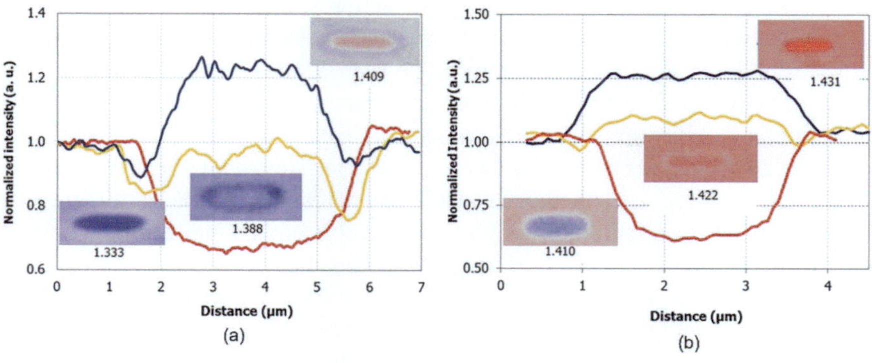

Fig. 8.4 Pixel intensity analysis of **a** *E. coli* and **b** *S. flexneri*. Reproduced from [24] with permission from the Royal Society of Chemistry

subjected to tensile (negative pressure) conditions, where further pressure reduction leads to rupture and vapor cavity formation. This critical pressure threshold for cavitation initiation defines the tensile strength. However, existing measurement techniques have proven challenging to implement, with most experimental results showing significant deviations from theoretical predictions. The development of a simple yet precise method for quantifying water's tensile strength would provide valuable insights for both fundamental research and practical engineering applications. Moreover, such methodology could be extended to investigate analogous properties in other soft material systems.

The measurement of tensile strength traces back to Berthelot's early experiments, where water was stretched and ruptured in a sealed tube during cooling [26]. Subsequent studies employed various techniques to generate negative pressure for liquid rupture, including capillary tubes [27], shock reflectors [28, 29], and acoustic waves [30]. However, significant discrepancies exist between experimental results and theoretical values (approximately 130 MPa for water), primarily due to heterogeneous nucleation in large water volumes [31, 32]. Most experimental measurements fall below 30 MPa, with the mineral inclusion method being the sole exception. This approach demonstrates that microscopic aqueous inclusions in minerals can sustain tensions nearing 140 MPa at 42 °C [33], although it is with experimental complexities requiring high-temperature sample preparation (300–400 °C) and autoclave conditions for precise pressure–temperature control within inclusions. Moreover, tensile strength is indirectly derived from extrapolating equations of state, introducing simulation limitations. Microfluidics offers an alternative solution by enabling homogeneous nucleation through controlled water volumes [34]. While successful for measuring water and ethanol tensile strengths through bubble distribution analysis, this method is unsuitable for high-viscosity liquids like oil and glycerol, as simulations rely on inviscid flow assumptions.

This section presents an experimental method utilizing optofluidic techniques to directly measure the tensile strength of water. The approach involves stretching water through laser-induced shock wave interaction with an air–water interface in a microchannel. By analyzing cavitation nucleation statistics at varying distances between the laser spot and the interface, the reproducible cavitation threshold is identified as the tensile strength. Microscopic confinement minimizes heterogeneous nucleation probability and reduces experimental times to hundreds of nanoseconds. This innovative method eliminates reliance on complex simulations and cumbersome procedures, enabling investigation of tensile strength across various soft materials. Notably, the low-cost optofluidic chip enables precise temperature and fluid control while offering seamless integration into functional systems. Its portability makes it ideal for field-deployable applications requiring rapid sample testing.

The measurement principle for liquid tensile strength is schematically shown in Fig. 8.5. When a microfluidic channel is partially filled with liquid, the hydrophobic PDMS walls create a relatively flat air–liquid interface (Fig. 8.5a). A pulsed laser beam focused into the liquid medium ionizes liquid molecules within the focal volume, generating plasma. This plasma rapidly expands, forming both a spherical shock wave and a cavitation bubble upon recombination (Fig. 8.5a). As the bubble

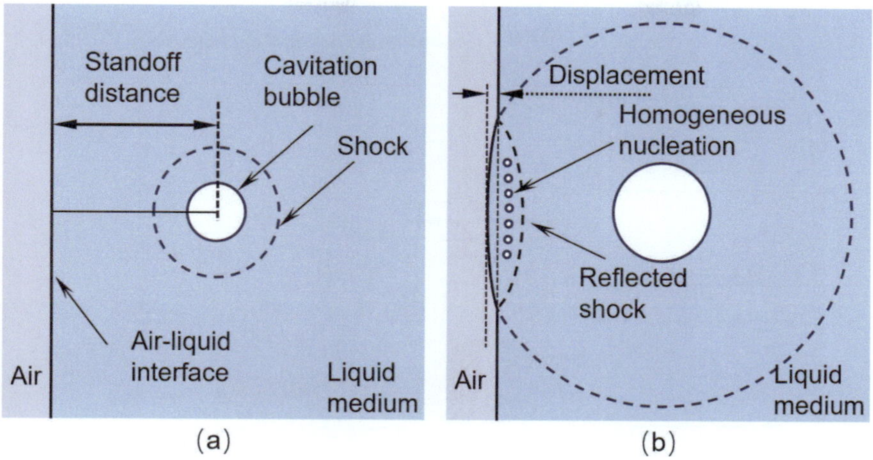

Fig. 8.5 Schematic illustration of the principle of tensile strength measurement of liquids

expands, the shock wave reaches the air–liquid interface. Due to acoustic impedance mismatch, this wave reflects as a tension wave at the interface (Fig. 8.5b). The reflected shock wave generates negative pressure at the air–water interface, which is precisely quantified by measuring both shock wave propagation and interface displacement. When this negative pressure exceeds the liquid's tensile strength, water rupture is observable through vapor bubble nucleation near the interface. By adjusting the standoff distance (distance between laser focus and interface), the tensile strength of water can be systematically determined.

The experimental setup for shock wave generation and imaging (Fig. 8.6) comprised a PDMS microchannel (400 μm width × 85 μm height) partially filled with air-saturated deionized water mounted on an inverted microscope. An Nd laser simultaneously delivered 7.5 ns pulses at 532 nm (green, via optical fiber for illumination) and 1064 nm (infrared, focused by a 40 × objective, NA 0.8) into the microchannel. Top-view images were captured by a CCD camera synchronized with the green pulse, while the temporal delay between pulses was measured using an ultrahigh-speed photodetector and recorded by a digital storage oscilloscope.

The temporal dynamics of laser-induced shock waves and their interaction with the air–water interface in the microchannel are presented in Fig. 8.7. With a near $-90°$ contact angle between water and the PDMS surface, the interface remains approximately flat. Upon laser triggering, optical breakdown occurs near the air–water interface at a standoff distance of 138.6 μm, generating a spherical shock wave from plasma expansion, which is identified as dark fringes in images. Following plasma recombination, a spherical bubble forms (evident at $t = 48.0$ ns, Fig. 8.7a). Upon reaching the interface (Fig. 8.7b), the shock wave reflects as a tensile wave due to acoustic impedance mismatch, creating a high-negative-pressure nucleation region (pseudo-colored yellow). Vapor bubbles, as indicators of water rupture, emerge here

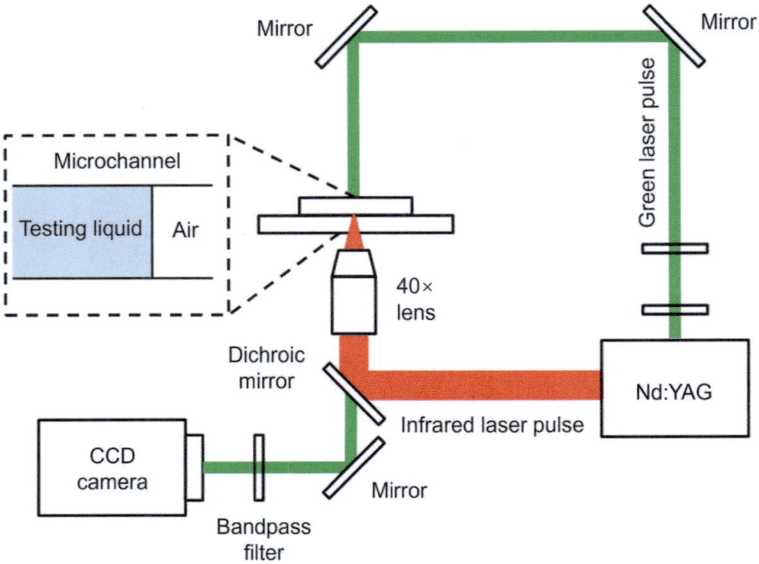

Fig. 8.6 Schematic illustration of the experimental setup for the tensile strength measurement of liquids

at $t = 97.9$ ns, with bubble count increasing as the reflected shock propagates (Fig. 8.7c, d).

When water is under tension, the air–water interface undergoes deformation, as illustrated in Fig. 8.7d. The displacement of the air–water interface (D) is defined as the distance between its original position and the point of maximum curvature on the deformed interface. The interface velocity (v) is calculated using

$$v = \frac{D}{t_r - t_c} = \frac{D}{t_n}, \tag{8.5}$$

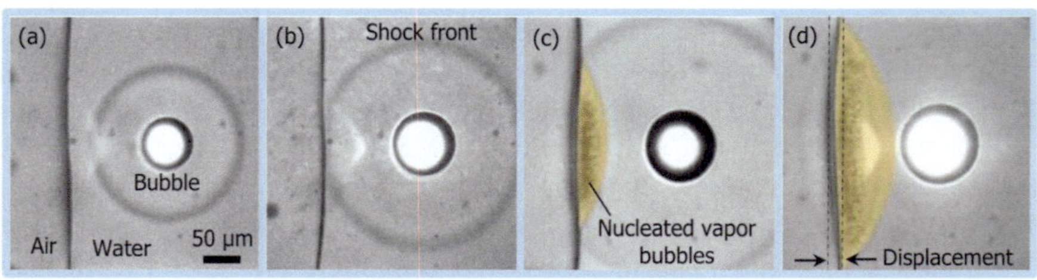

Fig. 8.7 Evolution of the shock interacting with the air–water interface: **a** 48.0, **b** 74.2, **c** 97.9 and **d** 108.1 ns after the laser shooting. Reproduced from [35] with permission from the Royal Society of Chemistry

where t_r is the image recording time, t_c is the shock compression time from the initial point to the interface, and t_n is the tension duration of the liquid in the nucleation region. The compression time depends on standoff distance and shock speed [35]. The shock pressure (p) is determined by

$$p = \frac{\rho u_s v}{2}, \tag{8.6}$$

where ρ is water density, and u_s is the shock speed.

Figure 8.8 demonstrates water rupture experiments conducted at six different standoff distances: 34.0, 51.1, 75.7, 117.8, 135.3, and 159.9 μm, with the nucleation region (bounded by the reflection shock and air–water interface) highlighted in yellow. To maintain a constant tension time ($t_n = 40$ ns), varying optical fiber lengths were employed for different standoff distances to accommodate the delayed green illumination pulse. The results reveal that shorter standoff distances yield greater air–water interface displacement, consistent with the anticipated rapid decay of shock pressure. The bright spot near the interface originates from the deformed polymer channel wall. By directly measuring the interface displacement from captured images at $t_n = 40$ ns, the interface velocities for all standoff distances were readily calculated using Eq. (8.5) and plotted in Fig. 8.9.

The shock speeds were measured across various standoff distances, ranging from 2.8 to 1.6 km/s. Using these values along with the interface velocities and water

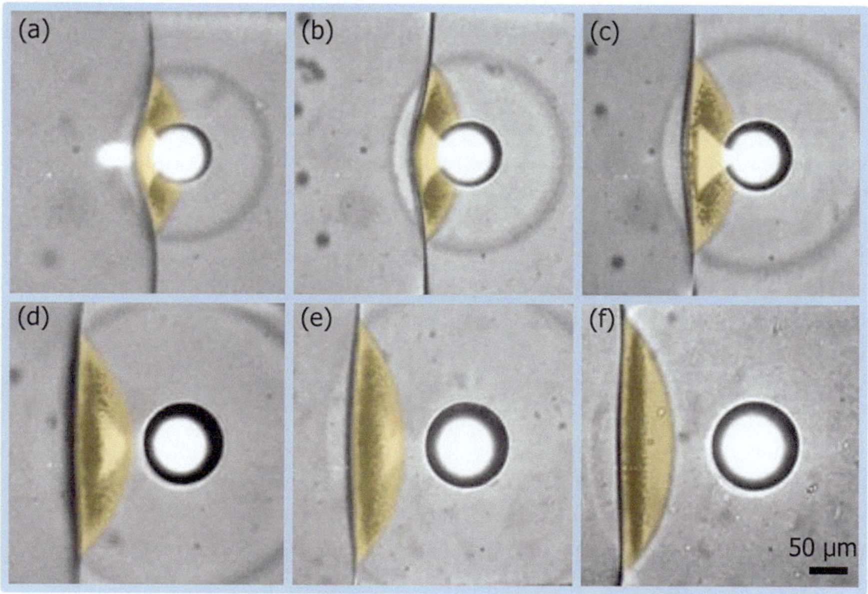

Fig. 8.8 Images of the nucleation induced by the shock reflection with the standoff distances of **a** 34.0, **b** 51.1, **c** 75.7, **d** 117.8, **e** 135.3 and **f** 159.9 μm. Reproduced from [35] with permission from the Royal Society of Chemistry

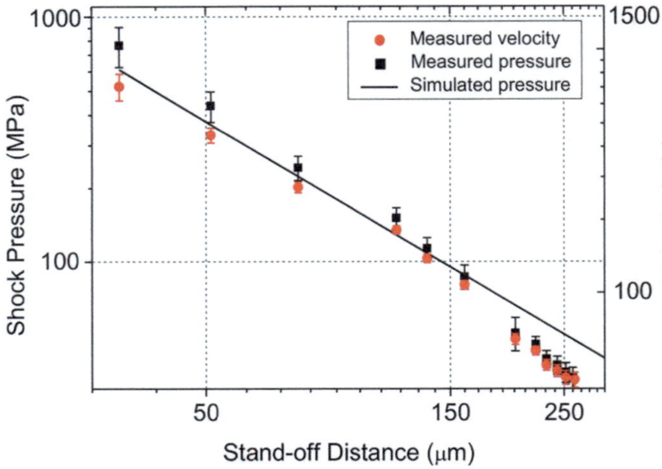

Fig. 8.9 Measured velocity of the interface and shock peak pressure values with various stand-off distances compared with the simulation results. Reproduced from [35] with permission from the Royal Society of Chemistry

density (998 kg/m^3 at 20 °C), Eq. (8.6) was employed to estimate the shock pressure. The experimental results were then compared with simulation data, as presented in Fig. 8.9. Within the standoff distance range of 30–300 μm, the shock pressure can be expressed as $P(r) = 5.0 \times 10^4\ r^{-1.25}$, where r denotes the standoff distance in micrometers. This relationship shows that the shock pressure decreases from 765 to 33 MPa as the standoff distance increases.

In the shock pressure measurement, the velocity of the deformed interface (v) was calculated as the average velocity over the entire tension time (t_n). Consequently, this approach yields higher values for shorter reflection times, leading to an underestimation of all experimental results in Fig. 8.9. Another critical factor is the presence of nucleated vapor bubbles, which overestimates the displacement (D). This effect becomes more pronounced with higher shock peak pressures. As a result, the shock pressure is particularly overestimated at shorter standoff distances. These two opposing factors partially cancel each other out, thereby reducing the overall measurement deviation.

The tensile strength of water is determined by measuring the shock pressure at a critical standoff distance, beyond which water rupture ceases. As demonstrated in Fig. 8.10a–e, homogeneous nucleation in the reflected shock region diminishes and ultimately vanishes when the standoff distance increases from 220.7 to 260.2 μm. Figure 8.10f further quantifies this trend by plotting the number of dark pixels in the nucleation region as a function of tensile pressure, reflecting cavitation yields. At the critical standoff distance of 260.2 μm, the corresponding shock speed and interface velocity are 1573.3 m/s and 42.4 m/s, respectively. The tensile strength is derived from the shock pressure calculated via Eq. (8.6), yielding a value of −33.3 ± 2.8 MPa for deionized water at room temperature (20 °C). This method can be

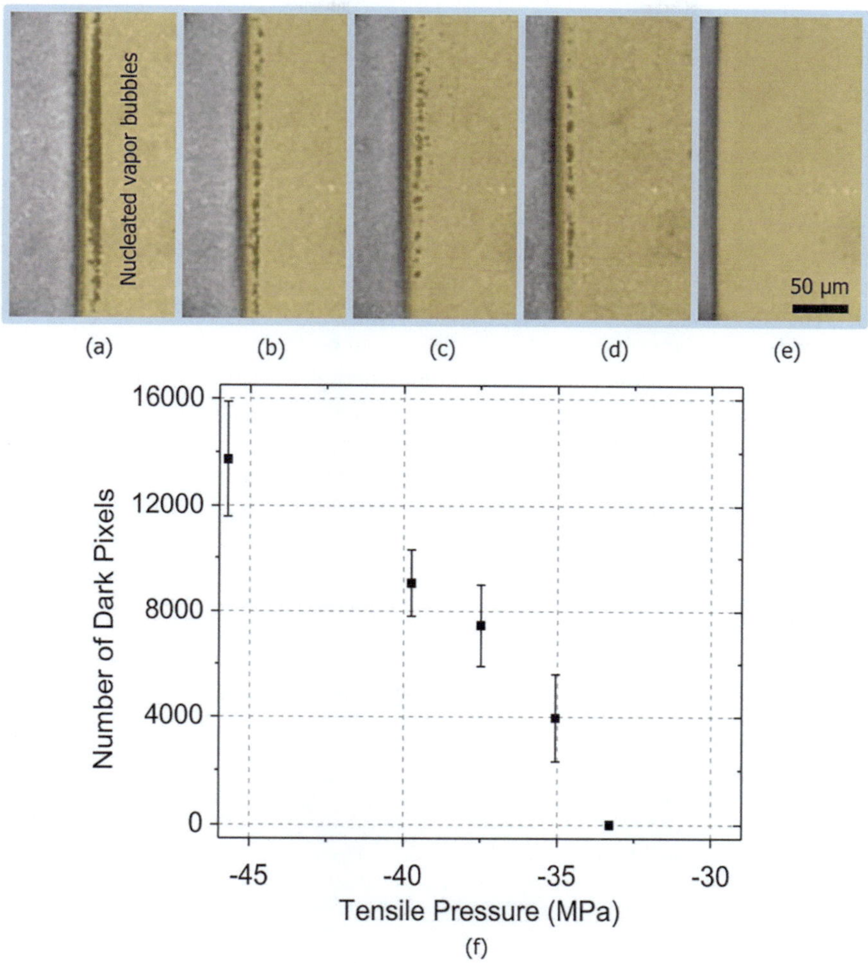

Fig. 8.10 Images of homogeneous nucleation with various standoff distances of **a** 220.7, **b** 231.0, **c** 242.4, **d** 252.0 and **e** 260.2 μm. **f** Number of dark pixels in the nucleation region as a function of the tensile pressure. Reproduced from [35] with permission from the Royal Society of Chemistry

adapted to measure the tensile strength of glycerol [35], demonstrating its versatility for determining the tensile strengths of diverse liquids.

8.4 Summary

Most techniques for the measurement can only get an average value of a big volume of samples, such as the average refractive index of suspended cells, which is lack of details and resolution. Moreover, experimental conditions have to be controlled and maintained throughout the measurements. It brings challenges to the reproducibility and accuracy of the measurement results. With the advanced development of microfluidic technique, sample volume is reduced, offering the possibility to study single cells and more stable experimental environment. For instance, on-chip measurement of tensile strength faces a low probability of heterogeneous nucleation. The integration of advanced optical imaging techniques into microfluidic platforms can also facilitate dynamic studies of samples. By combining with advanced optical imaging techniques to measure and monitor the refractive index distribution of cells, new biological insight will be obtained since localized changes down to 100 nm can be easily measured. The integration of microfluidics with optical measurement has advanced cutting-edge research toward high-throughput and precise control, enabled by optical imaging techniques offering high temporal, spatial, and axial resolution. The continuous efforts to improve its spatial and axial resolutions and extending its applications to water and environmental monitoring, food safety monitoring, other biomedical and biological areas such as virology and single molecule analysis will be the main future trends.

References

1. Barer R, Joseph S (1954) Refractometry of living cells part I. Basic principles. Q J Microsc Sci 95:399–423
2. Barer R, Ross KFA, Tkaczyk S (1953) Refractometry of living cells. Nature 171:720–724
3. Barer R (1957) Refractometry and interferometry of living cells. J Opt Soc Am 47:545–556
4. Barer R (1952) Interference microscopy and mass determination. Nature 169:366–367
5. Sha'afi RI, Rich GT, Sidel VW, Bossert W, Solomon AK (1967) The effect of the unstirred layer on human red cell water permeability. J Gener Physiol 50:1377–1399
6. Ghosh N, Buddhiwant P, Uppal A, Majumder SK, Patel HS, Gupta PK (2006) Simultaneous determination of size and refractive index of red blood cells by light scattering measurements. Appl Phys Lett 88:084101
7. Liang XJ, Liu AQ, Lim CS, Ayi TC, Yap PH (2007) Determining refractive index of single living cell using integrated microchip. Sens Actuators A 133:349–354
8. Cuerrier CM, Chabot V, Vigneux S, Aimez V, Escher E, Gobeil F Jr, Charette PG, Grandbois M (2008) Surface plasmon resonance monitoring of cell monolayer integrity: implication of signaling pathways involved in actin-driven morphological remodeling. Cell Mol Bioeng 1:229–239
9. Moh KJ, Yuan XC, Bu J, Zhu SW, Gao BZ (2008) Surface plasmon resonance imaging of cell-substrate contacts with radially polarized beams. Opt Express 16:20734–20740
10. Lee J, Lee C, Lin E, Wei P (2008) Single live cell refractometer using nanoparticle coated fiber tip. Appl Phys Lett 93:173110
11. Wyatt PJ (1970) Cell wall thickness, size distribution, refractive index ratio, and dry weight content of living bacteria (Staphylococcus aureus). Nature 226:277–279
12. Quist GM, Wyatt PJ (1985) Empirical solution to the inverse-scattering problem by the optical strip-map technique. J Opt Soc Am A 2:1979–1985
13. Flynn RA, Shao B, Chachisvilis M, Qzkan M, Esener SC (2005) Two-beam optical traps: refractive index and size measurements of microscale objects. Biomed Microdevice 7:93–97
14. Zhu H, White IM, Suter JD, Dale PS, Fan X (2007) Analysis of biomolecule detection with optofluidic ring resonator sensors. Opt Express 15:9139–9146

15. Arnold S, Khoshsima M, Teraoka I, Holler S, Vollmer F (2003) Shift of whispering-gallery modes in microspheres by protein absorption. Opt Lett 28:272–274
16. St-Gelais R, Masson J, Peter YA (2009) All-silicon integrated Fabry-Pérot cavity for volume refractive index measurement in microfluidic systems. Appl Phys Lett 94:243905
17. Domanchuk P, Littler ICM, Cronin-Golomb M, Eggleton BJ (2006) Compact resonant integrated microfluidic refractometer. Appl Phys Lett 88:093513
18. Song WZ, Zhang XM, Liu AQ, Lim CS, Yap PH, Hosseini HM (2006) Refractive index measurement of single living cells using on-chip Fabry-Pérot cavity. Appl Phys Lett 89:203901
19. Rappaz B, Marquet P, Cuche E, Emery Y, Depeursinge C, Magistretti PJ (2005) Measurement of the integral refractive index and dynamic morphometry of living cells with digital holographic microscopy. Opt Express 13:9361–9373
20. Charrière F, Marian A, Montfort F, Kuehn J, Colomb T (2006) Cell refractive index tomography by digital holographic microscopy. Opt Lett 31:178–180
21. Ikeda T, Popescu G, Dasari RR, Feld MS (2005) Hilbert phase microscopy for investigating fast dynamics in transparent systems. Opt Lett 30:1165–1167
22. Dirckx JJJ, Kuypers LC, Decraemer WF (2005) Refractive index of tissue measured with confocal microscopy. J Biomed Opt 10:044104
23. Curl CL, Bellair CJ, Harris T, Allman BE, Harris PJ, Stewart AG, Roberts A, Nugent KA, Delbridge LMD (2005) Refractive index measurement in viable cells using quantitative phase-amplitude microscopy and confocal microscopy. Cytometry A 65A:88–92
24. Liu Y, Chin LK, Ser W, Ayi TC, Yap PH, Bourouina T, Leprince-Wang Y (2014) An optofluidic imaging to measure the biophysical signature of single waterborne bacterium. Lab Chip 14:4237–4243
25. Bryant FD, Seiber BA, Latimer P (1969) Absolute optical cross sections of cells and chloroplasts. Arch Biochem Biophys 135:97–108
26. Berthelot M (1850) Sur quelques phenomenes de dilatation forcee des liquides. Annales de chimie et de physique 30:232–237
27. Briggs LJ (1950) Limiting negative pressure of water. J Appl Phys 21:721–722
28. Wilson DA, Hoyt JW, Mckune JW (1975) Measurement of tensile strength of liquids by an explosion technique. Nature 253:723–725
29. Williams PR, Williams RL (2004) Cavitation and the tensile strength of liquids under dynamic stressing. Mol Phys 102:2091–2102
30. Herbert E, Balibar S, Caupin F (2006) Cavitation pressure in water. Phys Rev E 74:041603
31. Fisher JC (1948) The fracture of liquids. J Appl Phys 19:1062–1067
32. Blander M, Kartz J (1975) Bubble nucleation in liquids. AIChE J 21:833–848
33. Zheng Q, Durben DJ, Wolf GH, Angell CA (1991) Liquids at large negative pressure: water at the homogeneous nucleation limit. Science 254:829–832
34. Ando K, Liu AQ, Ohl CD (2012) Homogeneous nucleation in water in microfluidic channels. Phys Rev Lett 109:044501
35. Li ZG, Xiong S, Chin LK, Ando K, Zhang JB, Liu AQ (2015) Water's tensile strength measured using an optofluidic chip. 15:2158–2161

Chapter 9
Optofluidic Devices

9.1 Introduction

Based on the optofluidic components described previously, diverse detection and analytical platforms have been developed for applications spanning biomedical research, biochemistry, pharmaceuticals, healthcare, and environmental monitoring. Conventional diagnostic procedures, however, often remain confined to centralized laboratories, requiring high cost and long turn-around time. Some analyses even need to culture the cells or transport large molecules into the cells, which require skilled technical personnel and are limited with low throughput. Optofluidics addresses these limitations by integrating optical and microfluidic functionalities within miniaturized Lab-on-a-Chip (LoC) devices. The integration features of optofluidic devices can accelerate the reaction speed due to the short diffusion distance, high surface-to-volume ratio, and efficient heat transfer. It opened up a platform to realize precise manipulation of fluid at sub-micron scale and bioparticles involving trapping, sorting, and separation etc. Sensitivity is improved due to the efficient mass transfer in a small volume and enhanced light–matter interaction using on-chip optical modules. Integrating optofluidic components onto a microchip can substantially cut fabrication costs and dramatically miniaturize the entire system. It is promising and possible to make the system portable for field tests even in remote regions that cannot afford high-cost equipment. Those advantages make optofluidics a promising solution for bio-diagnosis. The advances of optofluidic technology will be beneficial for addressing global health issues, such as the COVID-19 pandemic, by utilizing optofluidic cell analysis platform. This chapter discusses emerging optofluidic biodiagnostic devices, focusing on three key functionalities: refractometry, cell poration, and virus detection.

S. Xiong and A.-Q. Liu, *Optofluidics: Fundamentals and Applications*,
Microfluidics and Nanophotonics: Science and Engineering 4,
https://doi.org/10.1007/978-981-95-6296-1_9

9.2 Optofluidic Fabry-Pérot Refractometers

This section details two optofluidic Fabry-Pérot (FP) resonator configurations: (1) an optofluidic microlens FP resonator and (2) a complementary optofluidic FP resonant system. The first design incorporates a microchannel-embedded microcavity for cellular manipulation, coupled with dynamically tunable optofluidic microlenses that enable real-time optical coupling optimization. The second system implements a complementary FP architecture specifically designed for single-cell refractive index quantification, utilizing a precision single-wavelength intensity measurement methodology for live cell analysis.

The requisite precision for cellular refractometry varies significantly based on the biological context. Clinically, a refractive index precision of 0.005 is sufficient to distinguish normal red blood cells from those in anemia cases [1], while and a precision of 0.02 enables differentiation between G_2/M and G_0/G_1 cells in the cell cycle [2]. However, detecting minute refractive index shifts in single cells, such as those induced by antigen binding, demands a precision smaller than 0.001 [3]. To meet these stringent requirements, a high-finesse FP refractometer with enhanced precision and sensitivity is proposed. Figure 9.1a illustrates this optofluidic microlens FP refractometer [4]. The optofluidic microlens FP refractometer comprises a central microchannel functioning as the FP cavity, flanked by a pair of expansion chambers each featuring two inlets for liquid microlens formation via controlled merging of three laminar flows. Integrated optical fibers include one pair for optical input/output and a second pair dedicated to sample optical trapping. Microlens curvature is dynamically tuned within the expansion chambers by adjusting the relative flow rates of the constituent fluids. As shown in Fig. 9.1b, the sidewalls of the microchannel are gold-coated to maximize reflectance, essential for achieving high resonator finesse. For cellular measurements, a single cell is optically trapped within the cavity, and a differential measurement scheme simultaneously quantifies its effective refractive index and diameter.

For a planar FP cavity with reflectance of R, the transmission function is given as

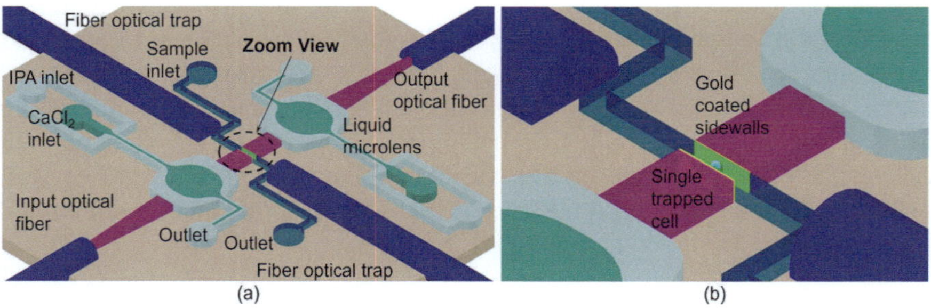

Fig. 9.1 Schematic illustrations of **a** the optofluidic microlens FP refractometer, and **b** single cell optically trapped in the FP cavity

$$T = \frac{1}{1 + F\sin^2[(\pi/\lambda)2nL]}, \quad (9.1)$$

where λ is the wavelength of the incident light, n is the refractive index of the medium in the cavity, L is the cavity length and F is the coefficient of finesse defined as

$$F = \frac{\pi\sqrt{R}}{1 - R}. \quad (9.2)$$

High-finesse resonators exhibit sharper resonant peaks and lower transmission minima in their spectra, enhancing refractive index measurement precision by enabling more accurate determination of resonant wavelength positions. To achieve a precision of 0.0005, the cavity finesse has to be at least 10. Key factors governing finesse include mirror reflectance, input light characteristics (e.g., coherence), and propagation loss. While the highest reflectance contrast occurs with an air-filled cavity, filling it with a higher refractive index solution reduces this contrast and thus the reflectance. To compensate and boost reflectance at the PDMS-liquid interfaces, a thin gold layer is sputter-coated on the microchannel sidewalls. This gold coating achieves an interface reflectance of 90–97% [5], critically dependent on gold layer thickness and PDMS sidewall smoothness. The resultant enhanced reflectance is essential for realizing the necessary high finesse in this optofluidic FP resonator.

Absorption and scattering losses within the cavity medium degrade resonator finesse. Mitigating these losses requires optimizing three key parameters: (1) selecting a medium with lower absorption coefficient, (2) minimizing propagation length, and (3) operating at wavelengths exhibiting lower absorption. With the involvement of liquid in the optofluidic FP systems, wavelengths near 1550 nm must be avoided due to high water absorption [6]. Furthermore, PDMS transparency imposes additional wavelength constraints. PDMS exhibits negligible absorption between 400–800 nm and 1250–1350 nm [7], defining the suitable operating wavelength ranges for precise refractive index measurement within this optofluidic FP refractometer.

Achieving high resonator finesse necessitates collimating the light emerging from the optical fiber. While various collimation techniques exist, such as packaged aspherical lenses, hollow-core waveguides, optical fiber slabs, capillary tubing, and epoxy microlenses, each presents challenges for integrated on-chip implementation. Aspherical lenses are typically suited for free-space optics. Integrating waveguides or fiber slabs is costly and complex. Although epoxy microlenses offer better potential for single-chip integration, their fabrication lacks flexibility: precise control over microlens curvature and height is difficult. Furthermore, using the fiber facets themselves as resonator mirrors introduces significant drawbacks; fiber misalignment and poor facet quality directly degrade resonator finesse and quality factor.

The spherical Fabry–Perot (FP) resonator offers superior light manipulation and confinement, presenting a more stable configuration inherently insensitive to geometrical misalignment (Fig. 9.2a). Its spherical curvatures perform two key functions: collimating diverging light and confining collimated light for resonance between highly reflective surfaces. Building on this concept, the proposed optofluidic microlens FP resonator utilizes a pair of high-performance liquid microlenses (Fig. 9.2b) to achieve real-time tunable light collimation. A gold-coated microchannel, positioned between these microlenses and pre-aligned during fabrication to prevent misalignment losses, serves as the resonant cavity, enabling both light confinement/resonance and single-cell manipulation. Crucially, this design decouples light collimation from the resonant cavity, providing greater flexibility in designing sample manipulation strategies.

The optofluidic FP refractometers are fabricated primarily using polymer soft lithography. This process achieves critical dimensional precision: angular imperfections measure less than 0.1°, and sidewall roughness is below 100 nm. Given that this roughness is significantly smaller than the incident light wavelength (1275 nm), reflections from the sidewalls are predominantly specular. Consequently, sidewall surface roughness has a negligible impact on the resonant cavity performance. For resonator characterization, a superluminescent diode (SLED) with a central wavelength of 1275 nm and bandwidth of 70 nm serves as the light source. Transmission spectra are detected using an optical spectrum analyzer with a resolution of 0.01 nm. Liquids are introduced into the chip through syringe pumps. Fluorescein dye is added to the flows to visualize the light path, and the device is imaged using a microscope

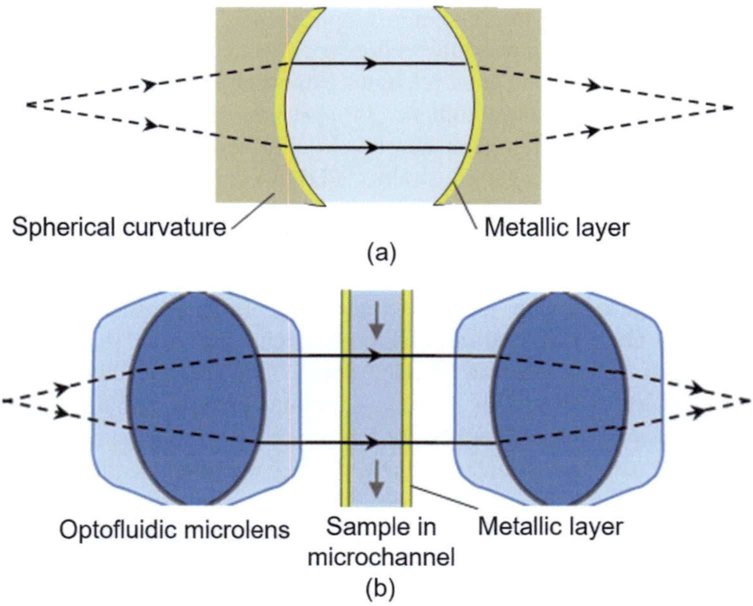

Fig. 9.2 Schematic illustrations of **a** the physics model of spherical FP resonator that consists of a pair of spherical curvature coated with metallic layer, and **b** optofluidic microlens FP resonator with a pair of liquid microlenses for light confinement. Reproduced from [4] with the permission of AIP Publishing

equipped with a CCD camera. Additionally, a high-power 1550 nm laser (maximum power 45 mW) is employed to implement the fiber optical trap.

Utilizing three laminar flow streams within an expansion chamber enables the dynamic formation of liquid microlenses, offering tunable control over both the radius of curvature and lens shape, as detailed in Chap. 2. The biconvex configuration provides superior focusing power or shorter focus distance compared to other microlens types. This allows light focusing or collimation within a significantly reduced spatial footprint, enabling more compact optical systems and smaller chips. In addition, the formation of biconvex microlens simplifies operation and enhances robustness, because both cladding streams share a single inlet and syringe pump, ensuring inherently matched flow rates. For the liquid microlenses, a 4.16 mol/L calcium chloride solution ($CaCl_2$, $n = 1.462$) serves as the core fluid, while isopropanol (IPA, $n = 1.382$) is the cladding fluid.

The finesse of the FP cavity exhibits a significant enhancement from 3.48 to 18.79 upon forming liquid microlenses with a 138 μm radius of curvature. Under this condition, the majority of incident light is collimated, facilitating efficient resonance and coupling back into the detection fiber. However, further reduction in the microlens curvature radius leads to light convergence from the optical fiber. Such converging light demonstrates poor resonance efficiency, consequently degrading the finesse (Fig. 9.3). Experimental results indicate that maintaining the curvature radius within 130–145 μm ensures a cavity finesse exceeding 15, defining the operational range for optimal performance.

To demonstrate the significance of achieving higher FP cavity finesse, measurement precision was investigated using cavities of differing finesse. With the optical spectrum analyzer (OSA) resolution fixed at 0.02 nm, 30 output spectra were captured for each cavity condition, and a spectral peak was located. Standard deviation and standard error were calculated from these 30 spectra. The low-finesse cavity (finesse = 3.48) yielded a standard deviation of 2.28×10^{-2} and a standard error of 4.0×10^{-3}. In contrast, the high-finesse cavity reduced these values to 2.7×10^{-3} and

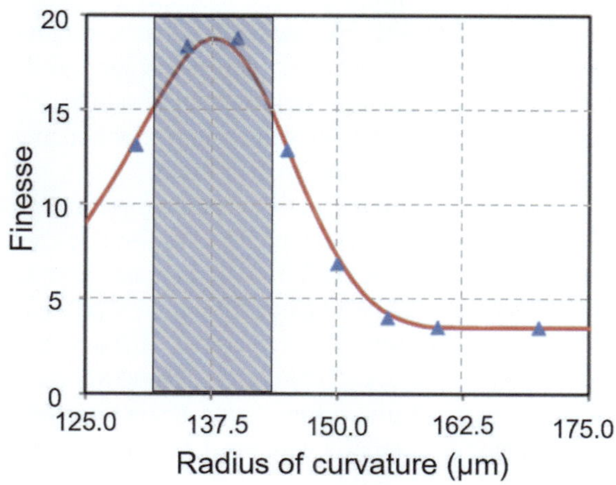

Fig. 9.3 Finesse of the resonant cavity versus the radius of curvature of the microlenses. The region of operation is highlighted in which the finesse is higher than 15. Reproduced from [4] with the permission of AIP Publishing

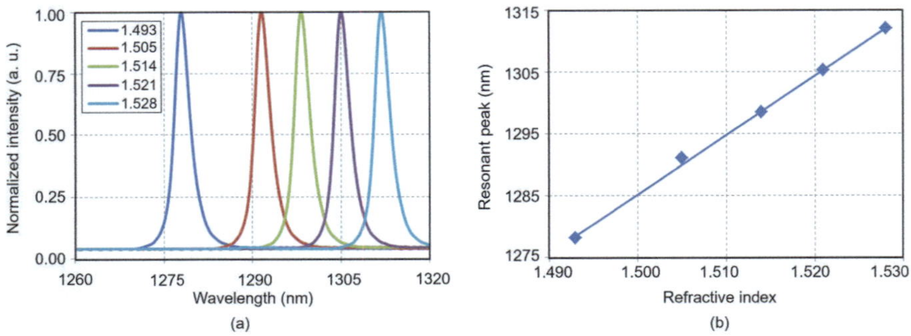

Fig. 9.4 **a** Transmission spectra of the FP resonator filled with buffer solutions of different refractive indices. **b** Shift of the resonant peak as a function of the refractive index of the buffer solution. Reproduced from [4] with the permission of AIP Publishing

8.0×10^{-4}, respectively. These results confirm that adopting a high-finesse resonant cavity substantially enhances measurement precision.

The optofluidic FP resonator was characterized as a volumetric refractometer. For system characterization, the FP cavity microchannel was filled with various liquids to quantify its refractive index responsivity, measured by resonant peak shifts. Certified refractive index liquids (Cargille Laboratories, series AA) covering a range of 1.493 to 1.528 were utilized for calibration. Figure 9.4a shows the output transmission spectra recorded as certified refractive index liquids are infused into the microchannel. The spectra reveal a distinct redshift of the resonant peak with increasing liquid refractive index. Figure 9.4b plots this resonant wavelength shift against refractive index, demonstrating the FP resonator's performance as a refractometer with a sensitivity of 960 nm/RIU and a detection range of 0.043 RIU.

Given the optical spectrum analyzer's highest resolution of 0.01 nm, the system achieves a refractive index resolution (smallest detectable change) of 1×10^{-5} RIU. The detection range is constrained by the cavity width since refractive index measurements are limited to one free spectral range (FSR). Wider cavities reduce the FSR, consequently narrowing the detection range. While cavity width governs the detection range, sensitivity depends solely on the wavelength of the detection light: longer wavelengths yield greater sensitivity. Using longer wavelengths also increases the FSR. However, wavelength selection must account for liquid absorption characteristics, as noted in the discussion.

The optofluidic FP resonator serves as a volumetric refractometer for quantifying concentrations in both calcium chloride solutions and fetal bovine serum samples. Experimental data (Table 9.1) demonstrates comparative performance against a commercial handheld refractometer with resolution of 1×10^{-4} RIU.

The optofluidic FP refractometer is redesigned for cellular measurements. To accommodate typical cell diameters (several microns to <20 μm), the microchannel width is set to 20 μm, necessitating beam waist reduction. While light from the single-mode fiber initially exhibits a 13 μm beam waist, divergence within the FP cavity expands it to ~100 μm. To counter this, a 15-μm aperture that fabricated as

Table 9.1 Measurement results of different buffer solutions in room temperature (20 °C)

	FP volume refractometer	Handheld refractometer
Calcium chloride solution ($CaCl_2 \cdot 2H_2O$)		
0.5 mol L^{-1}	1.34662 ± 0.00002	1.3466
1.0 mol L^{-1}	1.35441 ± 0.00003	1.3544
1.5 mol L^{-1}	1.36223 ± 0.00002	1.3622
2.0 mol L^{-1}	1.34988 ± 0.00002	1.3699
Fetal bovine serum solution (volume percentage)		
10%	1.33708 ± 0.00003	1.3371
20%	1.33870 ± 0.00002	1.3387
30%	1.34134 ± 0.00003	1.3413
40%	1.34321 ± 0.00002	1.3432

a gap between black-ink-filled microchannels is integrated between the input fiber and first expansion chamber. This aperture selectively transmits light from the input fiber while absorbing stray light in adjacent ink-filled channels. As a result, the collimated beam width reduces from 100 to 15 μm. This modification achieves an optimal resonator finesse of 13.31. Although the signal-to-noise ratio decreases 20-fold, resonance peaks remain clearly resolvable for measurements despite potential precision impacts.

The aperture-integrated FP resonator was characterized using certified refractive index liquids spanning 1.493 to 1.528 RIU. This yielded a detection sensitivity of 950 nm/RIU and a detection range of 0.043 RIU. Following characterization, the system measured Madin-Darby Canine Kidney (MDCK) cells. Cells were injected into the central microchannel filled with 1× phosphate buffer saline (PBS) solution and optically trapped between two fibers within the FP cavity, as shown in Fig. 9.5a. Optical forces were carefully adjusted to maintain spherical cell morphology without distortion, as shown in Fig. 9.5b. Flow was halted during trapping to eliminate hydrodynamic effects. To minimize phototoxicity, trapping duration was limited to <5 s, with spectra acquired after peak stabilization. The original PBS buffer was then replaced with a higher-concentration PBS solution. The refractive indices of the buffers before/after mixing were determined as $n_1 = 1.3553$ and $n_2 = 1.3611$, respectively. Four replicate measurements were conducted on two distinct MDCK cells using the same chip. The first cell has an effective refractive index of 1.3829 ± 0.0006 and size of 17.0 ± 0.5 μm, while the second measured cell has an effective refractive index of 1.3833 ± 0.0004 and size of 16.7 ± 0.3 μm. The experimental data demonstrated a measurement consistency with standard deviation of 5×10^{-4} RIU across replicates. The precision (equivalent to <0.05% variation) confirms the refractometer's capability to resolve subtle intracellular refractive index fluctuations, establishing its utility for single cell studies. Trypan blue viability testing (sample size: 30 cells) confirmed 90% cell survival post-measurement, indicating no significant procedural impact on cell health.

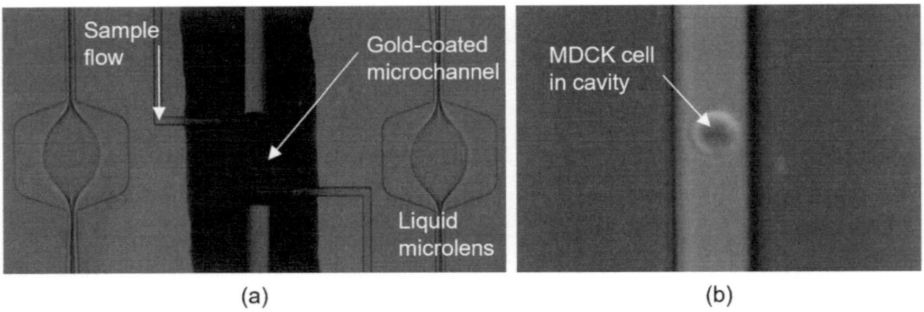

Fig. 9.5 Micrograph of **a** the FP resonator integrated with liquid microlenses for light collimation and optical fibers for optical trapping, and **b** a single MDCK cell being trapped in the cavity

Complementary optofluidic FP refractometer

The previous reported refractometers are based on the shift of the resonant peaks in the wavelength domain. Alternatively, refractive index can be quantified using a single-wavelength intensity modulation scheme in resonant microcavities, where particle or cell transit induces intensity fluctuations. By analyzing the intensity variation, the refractive index of the sample can be determined. Based on this working principle, a complementary optofluidic FP resonant system has been developed.

The complementary optofluidic FP resonant system integrates onto a chip featuring one inlet and bifurcated sample/waste outlets, as shown in Fig. 9.6a. Cell suspensions flow through a measurement region (Region b) containing three optical fiber pairs: bare Fiber-Pair I quantifying absorption/scattering attenuation, and gold-coated Fiber-Pairs A/B forming FP resonators to detect resonant mode intensity variations (Fig. 9.6b). Spherical cell sizing derives from temporal analysis of these signals, enabling refractive index calculation through FP intensity correlations. After the measurement region, a pressure-actuated valve controls flow routing as shown in Fig. 9.6c. Due to the presence of the restricting microstructures, the sample flow defaults to the waste outlet until cell detection triggers valve closure, diverting flow to the sample outlet. This configuration enables sensitive detection of rare cells with preconcentration capability through dynamic flow control.

Figure 9.7a shows a single FP resonator with gold coated facets. For facet reflectance R, the transmission function of the FP cavity is given by

$$T = \frac{(1 - R)^2}{1 + R^2 - 2R\cos(\delta)}, \tag{9.3}$$

where the phase difference δ between successive reflections is defined as

$$\delta = \left(\frac{2\pi}{\lambda}\right) 2nL\cos\theta. \tag{9.4}$$

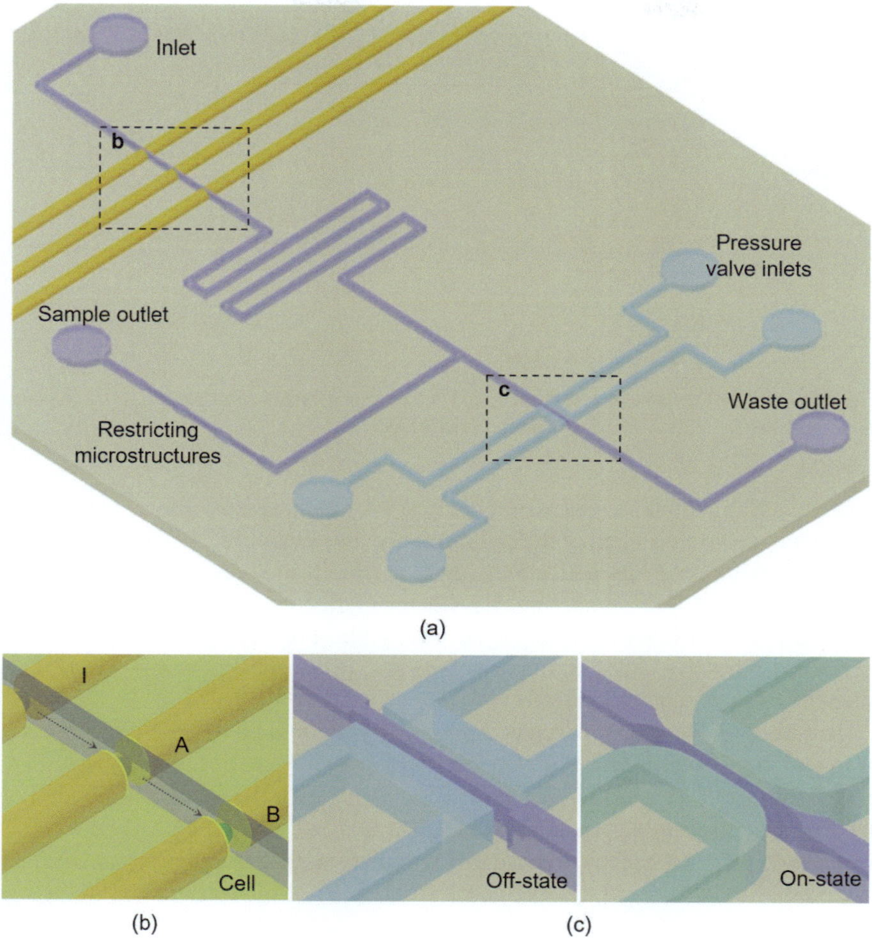

Fig. 9.6 Schematic illustrations of **a** the complementary optofluidic FP resonant system, **b** the optical measurement region, and **c** the pressure valve in the off- and on-states

Here n is the refractive index of the medium in the cavity, L is the cavity length,θ and λ represent the incident light angle and wavelength, respectively.

Figure 9.7b illustrates the variation of transmission intensity with different cell refractive indices in a 25-μm cavity. In the wavelength-based scheme, refractive index determination relies on resonant peak displacement within the transmission spectrum. Increasing the refractive index induces a redshift of the resonant wavelength (e.g., from 1.333 to 1.343). Conversely, the intensity-based scheme acquires transmission levels at fixed wavelengths (e.g., 514 nm).

Two critical constraints arise when using single FP resonators. First, measurement range is restricted to 0.05 RIU in intensity-based schemes due to symmetry in the transmission profile, merely half the range achievable with wavelength-based

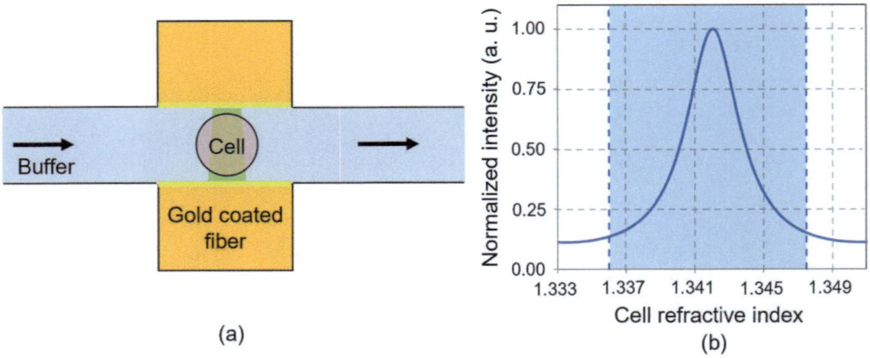

Fig. 9.7 **a** Single FP resonant cavity integrated with microchannel for sample injection, and **b** the intensity variation relative to different cell refractive indices

detection. This symmetry creates ambiguity at intermediate intensities. For example, one cannot determine whether the cell refractive index value is 1.340 or 1.344 when the transmission intensity is 0.5. Sensitivity is confined to specific refractive index regimes (1.336–1.347), rendering the system ineffective in low-gradient regions where differential intensity changes are negligible. For example, the system fails to resolve refractive index differences between 1.333 and 1.334 due to identical intensity responses at these values.

To address these limitations, a complementary optofluidic FP resonant system is proposed, as shown in Fig. 9.8a. Spectrally distinct FP cavities are precisely calibrated to operate in complementary sensitivity regimes (Fig. 9.8b). Resonator A has low sensitive regions ranging from 1.333 to 1.336 and 1.347 to 1.351, while resonator B provides enhanced sensitivity within these specific ranges. This configuration achieves two critical advancements: (1) elimination of measurement ambiguity through unique intensity-refractive index pairings, and (2) a two-fold extension of the dynamic range to 0.1 RIU. The system's expanded capability stems from each refractive index value generating distinguishable intensity responses across both resonators, effectively resolving previous single-resonator limitations.

Increased reflectance in FP resonators elevates cavity finesse, producing sharper resonant peaks with reduced bandwidth, which is a critical requirement for wavelength-based detection. However, for intensity-based measurements at fixed wavelength, high finesse is nonessential. Analysis confirms that while higher reflectance enhances intensity contrast between transmission maxima and minima, it proportionally narrows the sensitive measurement bandwidth. This establishes a fundamental trade-off between measurement sensitivity and operational range. Optimal performance (sensitivity $\approx$0.9, bandwidth $\approx$0.008 RIU) is achieved at $R = 0.4$–0.5. Critically, this moderate reflectance requirement eliminates the need for high-reflectance surfaces or high-finesse microcavities, significantly simplifying device fabrication.

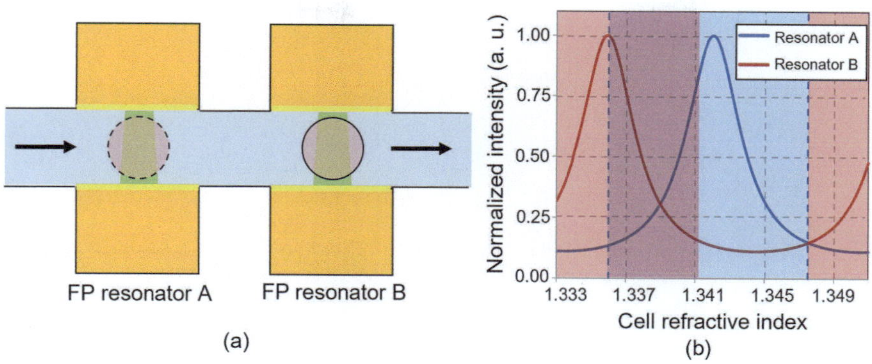

Fig. 9.8 **a** Complementary optofluidic FP resonant system, and **b** the intensity variation relative to different cell refractive indices

When measuring single cells with the complementary optofluidic FP resonant system, theoretical analysis identifies a key trade-off related to particle size. Smaller particles expand the measurement range (e.g., range doubles if particle size halves) because they cause smaller optical path changes for a given refractive index shift. However, this comes with reduced sensitivity. Importantly, cavity length influences neither the range nor the sensitivity. While wavelength-based schemes benefit from short cavities (producing larger, easier-to-detect shifts), these are difficult to fabricate. Fortunately, the single-wavelength intensity-based scheme performs well without requiring short cavities, thereby greatly simplifying the fabrication process.

The presence of a particle or cell within the cavity typically causes attenuation of the transmission intensity. This attenuation level depends on the particle/cell's absorption and scattering properties. To measure this effect, an additional pair of fibers without gold coating is integrated with the complementary optofluidic FP resonant system. These dedicated fibers quantify the intensity attenuation specific to different particles/cells. To isolate the refractive index information from the measured optical path difference, cell size is determined via signal modulation, as illustrated in Fig. 9.9. The time interval between Signal I and Signal A is denoted t_1, while the interval between signal A and signal B is t_2. Given the separation between Fiber-Pair I and FP resonator A is 800 μm, and between FP resonator A and FP resonator B is 400 μm, the flow velocity of the cell in the microchannel is calculated by

$$v_{\text{cell}} = \frac{1}{2}\left(\frac{800}{t_1} + \frac{400}{t_2}\right). \tag{9.5}$$

Using the derived flow velocity, the cell size is determined from the individual intensity signal. As the cell begins traversing the light beam, the transmission intensity decreases at time t_{bef}, as shown in Fig. 9.9b. When the cell fully exits the beam, the intensity recovers to its baseline level at time t_{aft}. Averaging the sizes obtained from all three fiber pairs yields the cell size according to

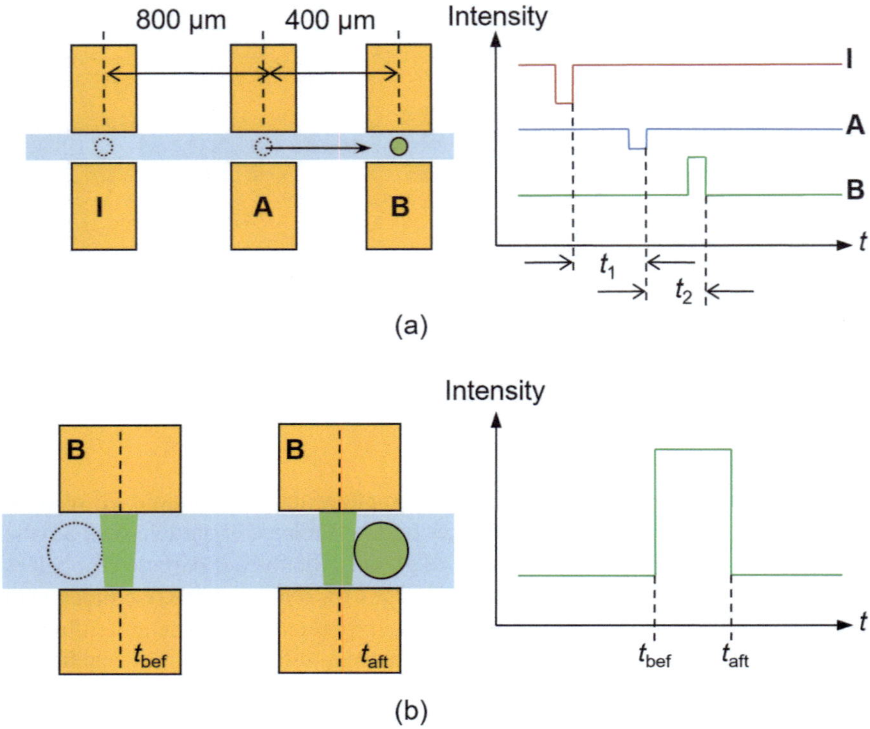

Fig. 9.9 Intensity signals observed with respect to time when single cell flows through **a** the measurement region, and **b** the FP resonator B

$$d_{\text{cell}} = -d_{\text{beam}} + \frac{1}{3} \sum_{n=3} v_{\text{cell}}(t_{\text{aft}} - t_{\text{bef}}), \tag{9.6}$$

where d_{beam} represents the width of the light beam.

The optofluidic FP refractometers are fabricated through polymer-based soft lithography techniques. As illustrated in Fig. 9.10a, the completed complementary optofluidic FP resonant system features integrated optical fibers secured by UV-curable epoxy at precisely controlled separation distances and optimal alignment. Figure 9.10b demonstrates the pressure valve in the off-state. When the valve switched to on-state, hydraulic pressure applied to side chambers induces microchannel wall deformation, effectively sealing the waste outlet to prevent liquid leakage, as shown in Fig. 9.10c.

The microfluidic system handles sample injection and pressure-driven valve operation. During valve control, DI water fills the microchannels initially, followed by computer-regulated pressure adjustment through the inlets. For optical measurements, fibers link to an external system with two modes. First, Transmission spectra of FP resonators are captured using a 633-nm He–Ne laser (30 nm bandwidth) input and spectrometer output. This setup is used for wavelength-based measurement scheme.

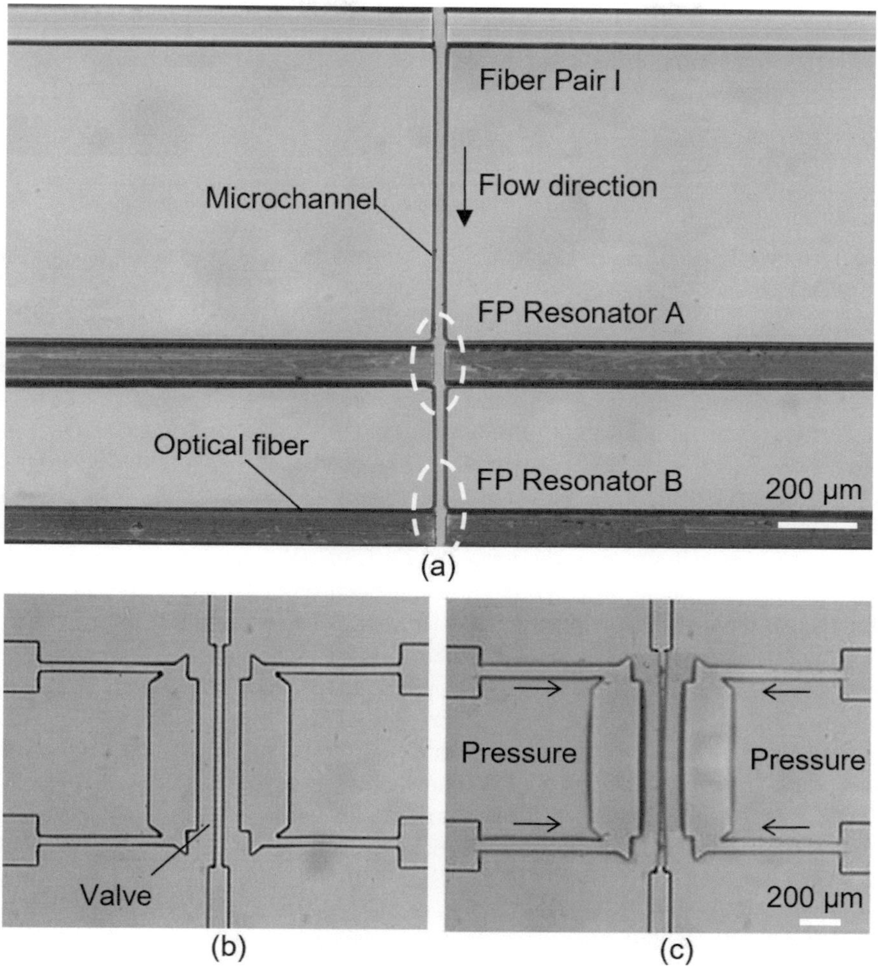

Fig. 9.10 Micrographs of **a** integrated fibers with desired separation distance and at optimal alignment, **b** the pressure valve in the off-state, and **c** the pressure valve in the on-state

Second, a 514-nm argon ion laser source splits light into three input fibers. Output intensities are recorded via PMT detectors optimized for 500 nm. This setup is used for single-wavelength intensity based measurement scheme.

The complementary optofluidic FP resonant system was employed to characterize MDCK cells suspended in PBS solution ($n = 1.356$). Precise unidimensional refractive index and size measurements critically depend on stable cellular flow patterns within the microchannel, as spinning or pulsation introduces significant error. High-speed camera monitoring confirmed that MDCK cells maintained a stable, rolling trajectory along the flow direction at constant velocity, as shown in Fig. 9.11. Based on the captured micrographs, the cell is rolling in the flow direction with a constant

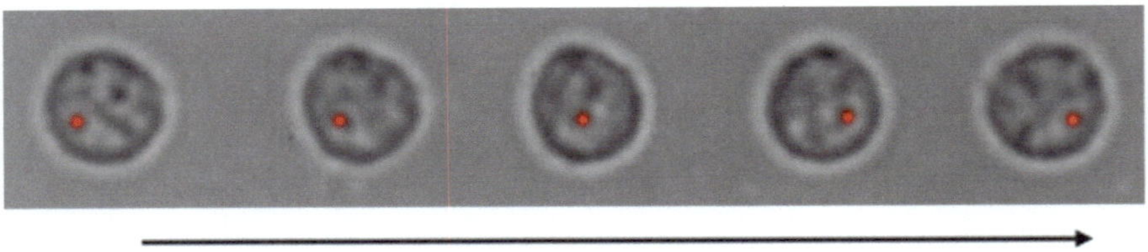

Fig. 9.11 Micrographs of the flowing MDCK cell in the microchannel. A point is highlighted in red to prominently show the rotating direction of the flowing cell

velocity. Furthermore, the spherical morphology of the MDCK cells minimizes the influence of flow dynamics on measurement accuracy.

The key measurement results are shown in Fig. 9.12. The measured MDCK cell size is 16.17 μm, with an attenuation/scattering coefficient (A_{AS}) of 0.9753. Shifts in FP resonance peaks (Fig. 9.12a) induced intensity variations (ΔI_{OPD}) of 0.3213 (resonator A) and 0.5147 (resonator B). The intensity variations of the FP resonant system for this cell size over the range 1.372–1.387 RIU is plotted in Fig. 9.12b. Based on the measured intensity variations, the cell refractive index is 1.382. Size and RI distributions for a population of 30 MDCK cells are presented in Fig. 9.12c, d, yielding a mean refractive index of 1.382 ± 0.001. To assess viability, cells were injected into the system, collected at the outlets, and stained with Trypan blue. The test revealed 96% cell viability (sample size: 50 cells).

Single-cell separation is not inherent in the design but is easily achievable via T-junction or hydrodynamic focusing structures constricting flow to single-cell width. These experiments utilized analogous methods to optimize single-cell measurement efficiency, with results confirmed microscopically. Cell stiction is negligible due to high flow velocity (~500 μm/s). Each cell is measured within 5 s, significantly improving throughput compared to previous systems. Measurement speed is limited primarily by data acquisition rate: slower flows yield higher resolution (more data points) at fixed acquisition rates. Sequential measurement is feasible, though cell sorting would require synchronized detection and flow switching. Additional functionalities can be easily integrated via this lab-on-a-chip platform.

9.3 Optofluidic Chip for Cell Poration

Previous studies on cavitation bubble dynamics demonstrate that laser-induced bubbles collapsing near rigid boundaries generate high-speed microjets [8, 9]. During collapse, bubbles elongate perpendicularly to the boundary, and asymmetric flows driven by secondary Bjerknes forces propel them toward the surface. This forms a boundary-directed microjet that penetrates the bubble, creating a toroidal structure. In this section, an array of cell trapping structures is used to immobilize individual cells while acting as rigid boundaries to initiate microjets for single-cell membrane

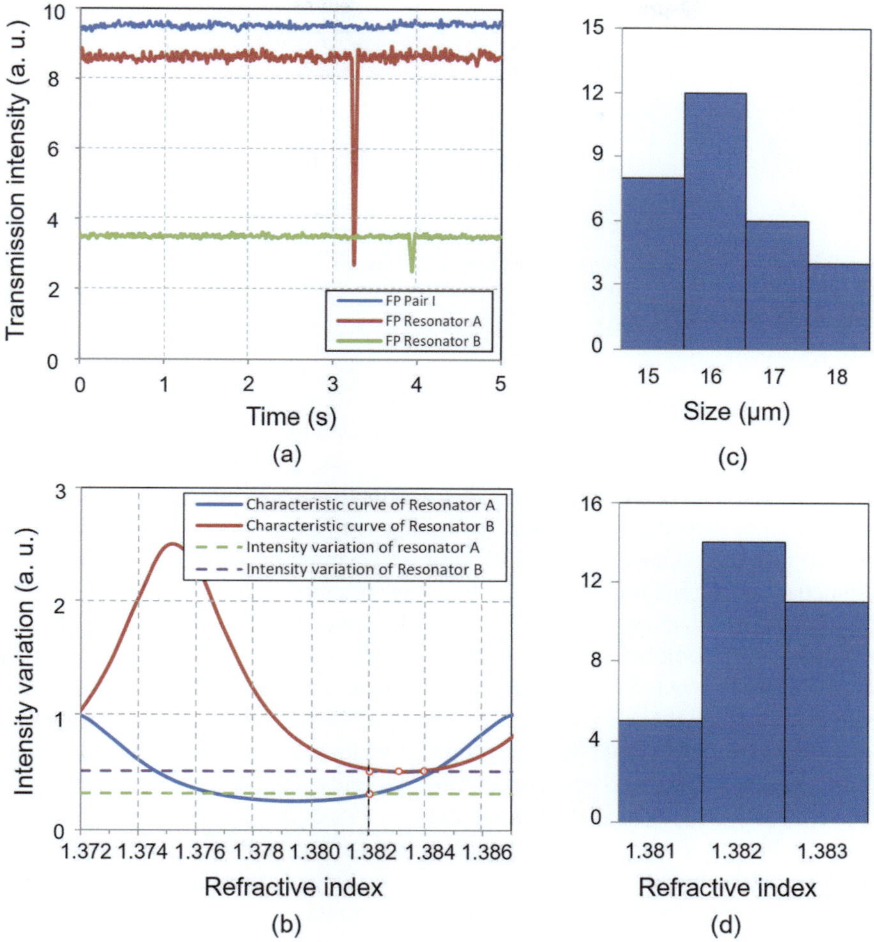

Fig. 9.12 **a** Measured intensity when the MDCK cell was flowing through the detection region. **b** Comparison with the measured intensity variations with the characteristic curves of both resonators in the refractive index range from 1.372 to 1.378 for a 16.17-μm cell. **c** Size and **d** refractive index distributions of 30 MDCK cells

poration [10]. Microjet strength is critically dependent on maximum bubble size and standoff distance, which is the distance between the bubble center and structure boundary. Optimization of these variables, particularly standoff distance, is essential for controlling pore size and maintaining cell viability.

The working mechanism of single-cell membrane poration begins with cell trapping in an equilateral triangular microchannel connected to a narrower rectangular channel (Fig. 9.13a). A 0.4% (w/v) saline trypan blue solution serves as extracellular medium, acting both as a poration biomarker and a laser absorber to facilitate bubble generation. A cavitation bubble is induced by focusing a 532 nm pulsed laser at a defined standoff distance (Fig. 9.13a). Due to the trapping structure, bubble expansion

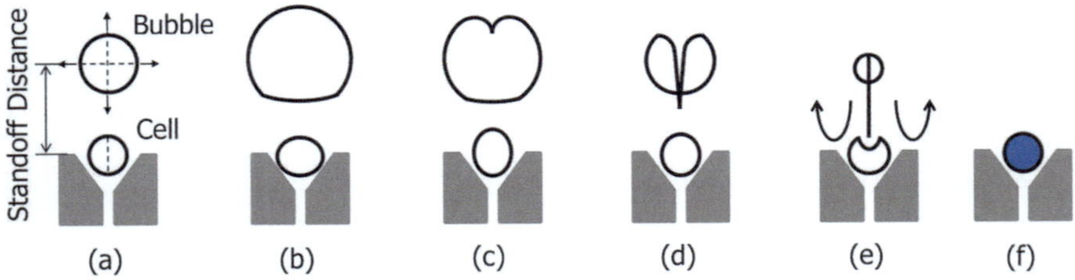

Fig. 9.13 Schematic illustration of the working mechanism of single-cell membrane poration using the cavitation bubble-induced microjet. Reproduced from [10] with permission from the Royal Society of Chemistry

and collapse are asymmetric (Fig. 9.13b, c). Bubble expansion compresses the cell against the trapping structure within 20 μs, followed by collapse-phase elongation toward the bubble and simultaneous formation of a cell-directed microjet (Fig. 9.13d) that evolves into counter-rotating vortices (Fig. 9.13e). This entire bubble dynamics sequence precedes millisecond-scale cell shape recovery and second-scale intracellular trypan blue uptake (Fig. 9.13f) with all asymmetric behaviors arising from geometric constraints imposed by the trapping structure.

Figure 9.14 illustrates the microfluidic chip design that incorporates an array of structures, each comprising an equilateral triangular converging channel (side length $w_1 = 20$ μm) connected to a rectangular channel of variable width ($w_2 = 5$–10 μm), with a uniform height of 27 μm. This geometry enables hydrodynamic cell trapping. When suspended cells enter the triangular region, they are arrested at the rectangular constriction (w_2), while subsequent cells flow around occupied traps, achieving high-efficiency single-cell isolation. Beyond immobilization, the asymmetric confinement directs bubble-induced microjets toward trapped cells, enabling integrated cell positioning and membrane poration. The design thus merges spatial control with physical poration capability within a single microarchitectural unit.

The microfluidic chip is fabricated in PDMS via soft lithography. The chip features converging trapping structures, with representative cell trapping shown in Fig. 9.15a. Myeloma cells (10^6 cells/ml) suspended in 0.4% trypan blue solution were perfused through the device at 0.5 μL/min. After complete chip filling, flow was terminated, resulting in trapping structures containing zero to multiple cells per site. As shown in Fig. 9.15b, statistical analysis across 94 structures confirmed single-cell occupancy is higher than 50%, demonstrating efficient isolation capability.

The single-cell membrane poration experimental system integrates optical, microscopic, and high-speed imaging components. A Q-switched Nd laser (532 nm wavelength, 6 ns pulse duration, 100 μJ energy) delivers pulses through an expanded beam path into an inverted microscope equipped with a 20× objective (NA = 0.75). This configuration focuses laser energy into the microfluidic chip's liquid medium. For bubble dynamics visualization, the microscope's condenser provides illumination while a dichroic mirror directs light to a high-speed CMOS camera (protected by a 532 nm notch filter). Two imaging protocols were employed: boundary interaction

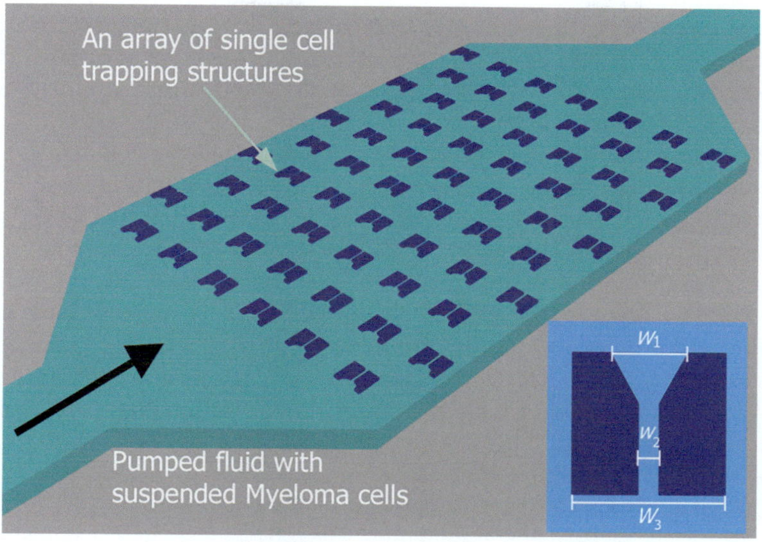

Fig. 9.14 Microfluidic chip schematic showing an array of single-cell trapping sites. Inset: Detailed view of a single trapping structure. Reproduced from [10] with permission from the Royal Society of Chemistry

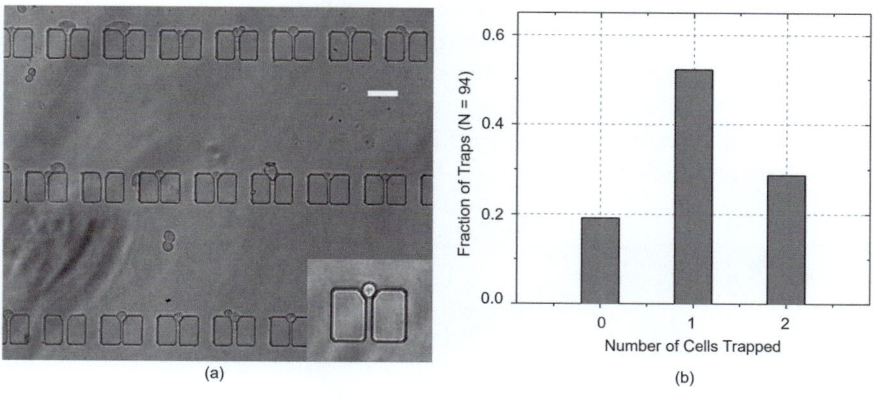

Fig. 9.15 **a** Micrograph of cell trapping. The scale bar is 50 μm. **b** Statistics of single cell trapping

studies at 300,000 fps (1 μs exposure), and cell deformation/poration recordings at 500,000 fps (370 ns exposure).

In the experiments, the laser is also used as a trigger signal for activating the high-speed camera. Due to the lack of synchronization between the camera and laser systems, the exact moment of bubble formation exhibits temporal variation within a single framing interval. At a recording rate of 300,000 frames per second, this timing uncertainty can extend up to 3 μs. Consequently, the initial frame may or may not capture the bubble formation. To establish a standardized reference point, $t = 0$ is chosen as either the instant when the bubble attains its maximum diameter, or the first appearance of the bubble in cases. The choice does not influence the experimental results but facilitates clearer interpretation of the data and more straightforward comparison with simulation outputs.

The experimental sequence records the complete spatiotemporal dynamics of a trapped myeloma cell undergoing laser-induced cavitation in trypan blue solution as shown in Fig. 9.16a. At $t = 0$ μs, laser focusing generates a cavitation bubble with a maximum diameter of 42 μm at 25 μm standoff distance, causing the cell to flatten horizontally against the trapping structure as the bubble expands. Subsequent bubble collapse at $t = 2$ μs triggers vertical cell elongation accompanied by clear bubble flattening, followed by asymmetric collapse ($t = 4$–6 μs) generating a microjet that produces distinct apical indentation. Bubble residue appears as a dark point at $t = 6$ μs. Cellular morphology gradually recovers over several milliseconds, returning to approximately original dimensions by $t = 8$ ms.

The bubble-cell interaction dynamics were numerically simulated using a potential flow solver, specifically an adapted boundary element method [11, 12] under axisymmetric and inviscid flow assumptions. The cell boundary was modelled as free fluid particles and the trapping structure as spatially fixed. Comparative analysis of experimental and simulation results (Fig. 9.16) shows that while both systems demonstrate cell stretching within the converging structure at $t = 0$ μs, the experimental cell exhibits less upper deformation than predicted numerically, attributed to unaccounted membrane tension and cytoplasmic viscoelasticity in the simulation. The simulation (Fig. 9.16b) provides enhanced temporal resolution of collapse dynamics (unresolvable by camera), revealing bubble shrinkage with major-axis perpendicular orientation at $t = 4.5$ μs, followed by center acceleration toward the structure that generates a high-speed microjet by $t = 5.2$ μs. This jetting flow correlates with the experimental indentation observed at $t = 6$ μs.

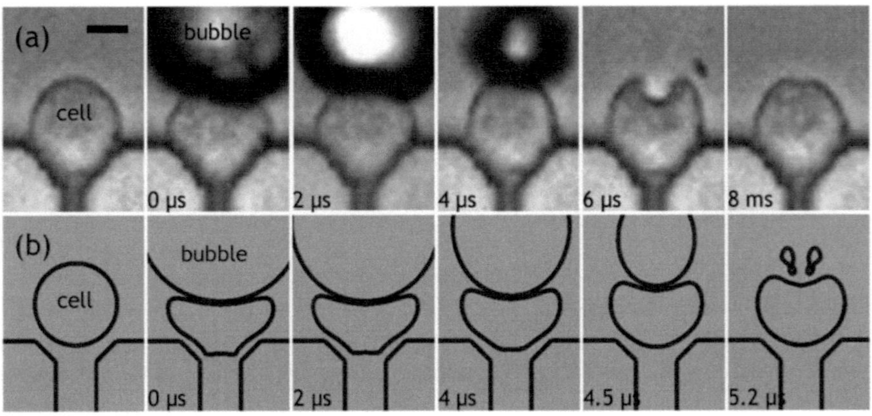

Fig. 9.16 **a** Experimental and **b** simulation results of the deformation of a single trapped cell from a cavitation bubble. The scale bar is 10 μm. Reproduced from [10] with permission from the Royal Society of Chemistry

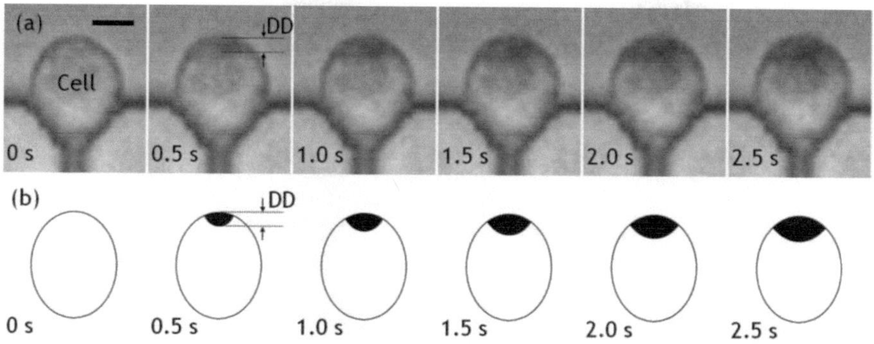

Fig. 9.17 **a** Selected micrographs of the trypan blue uptake process for the porated cell in 2.5 s. The scale bar is 10 μm. DD is defined as the distance between the diffusion frontier and the upper boundary of the cell. The diffusion frontier is defined as the 50% level between the background gray level value and the darkest gray level value. **b** Simulation results of the diffusion front. Reproduced from [10] with permission from the Royal Society of Chemistry

The study investigated the uptake mechanism of membrane-impermeable trypan blue dye following cavitation bubble collapse, as captured by high-speed imaging in Fig. 9.17a. Dye penetration occurs exclusively through the upper pole of the cell, gradually darkening the cytoplasm, with the grayscale change serving as the quantitative uptake metric. The observed dye front propagation rate (6 μm in 2.5 s) and temporal characteristics indicate purely diffusive transport dominates cytoplasmic staining.

To validate the experimental observations, the uptake dynamics were compared with solutions of the diffusion equation expressed as

$$\frac{\partial c(x, t)}{\partial t} = D\nabla^2 c(x, t), \tag{9.7}$$

where $c(x, t)$ represents trypan blue concentration (mol/m^3) and D is the diffusion constant. COMSOL Multiphysics was employed to model diffusion under key simplifications: identical time-independent diffusion coefficients intra/extracellularly, constant pore size throughout the process, and approximation of the cell as a 3D ellipsoid (major axis 12.5 μm, minor axis 10 μm) matching the initial morphology in Fig. 9.17a. The cell was centered in a domain matching experimental channel height, with isolation boundaries applied everywhere except at a single circular pore on the upper pole. Diffusion coefficients were defined as $D_{\text{out}} = 2.2 \times 10^{-10}$ m^2/s [13] for the extracellular domain and $D_{\text{in}} = 1.1 \times 10^{-12}$ m^2/s [14] for the intracellular compartment. Initial normalized dye concentrations were set at $C_{\text{out}} = 1$ and $C_{\text{in}} = 0$ for extracellular and intracellular domain, respectively. A fixed threshold ($C_{\text{threshold}} = 2.94 \times 10^{-4}$) demarcates the diffusional transport frontier. This computational framework enabled quantitative comparison of dye transport kinetics with experimental measurements.

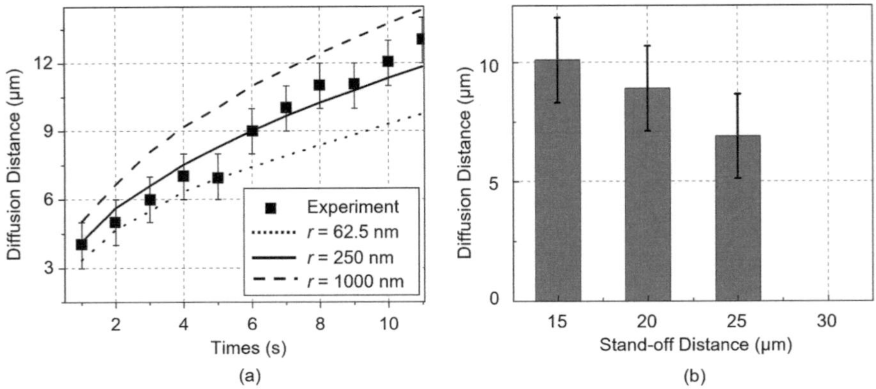

Fig. 9.18 **a** Comparison of simulated diffusion fronts for three distinct effective pore sizes with experimental data. **b** Averaged uptake distance of trypan blue into porated cells 3 s after the treatment and for different stand-off distances. The error bars depict the standard deviation for 10 repeated experiments. Reproduced from [10] with permission from the Royal Society of Chemistry

Figure 9.17b presents the simulated diffusion frontier for a pore radius of $r =$ 250 nm. Given the two adjustable parameters, $C_{\text{threshold}}$ and D, reasonable agreement with experimental data is achieved across pore sizes spanning 100 nm to 1 μm. As theoretically predicted, the frontier position exhibits characteristic $t^{1/2}$ dependence at shorter timescales as the solid line in Fig. 9.18a. Reduced pore radius ($r = 62.5$ nm, dotted line) decelerates diffusion kinetics, while a larger pore ($r = 1$ μm, dashed line) accelerates dye transport.

Results indicate that microjets generated by bubbles induce membrane strain, facilitating membrane poration and subsequent dye uptake. Standoff distance significantly influences poration efficacy. At larger distances (e.g., 30 μm), insufficient strain prevented membrane poration entirely. Experimental validation (Fig. 9.18b) quantifies this relationship through diffusion front progression 3 s post-bubble generation across four standoff distances (10 replicates each). Statistical analysis (*Student's* t-test, $p < 0.05$) reveals a progressive increase in mean diffusion distance as standoff distance decreases.

Figure 9.19 shows cellular dynamics during the first 35 s post-poration, analyzing both trypan blue uptake patterns and morphological recovery. The cytoplasmic staining intensity progressively increases as the dye penetrates deeper into the cell. Quantitative volume analysis reveals continuous cellular expansion, with the volume ratio V/V_0 reaching approximately 2.0 by $t = 35$ s. This swelling phenomenon is attributed to osmotic pressure gradients across the compromised membrane, where reduced structural resistance permits rapid fluid influx. Notably, membrane repair capacity appears dye-dependent, while the toxicity of trypan blue may impair recovery mechanisms, nontoxic compounds could theoretically avoid inducing such osmotic swelling.

Cell viability preservation constitutes a critical metric for evaluating membrane poration techniques. The inherent toxicity of trypan blue impedes accurate viability

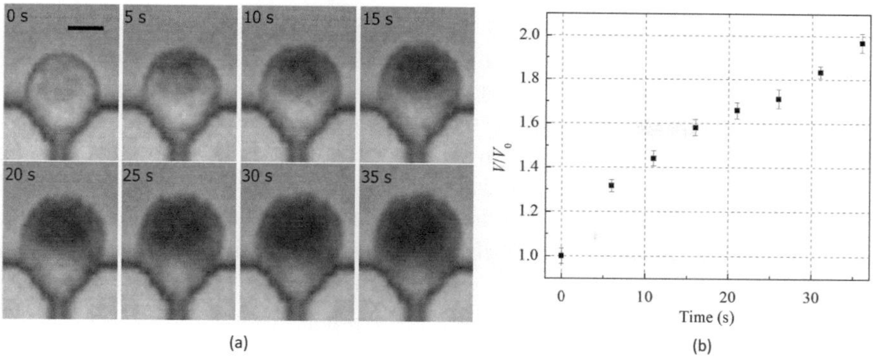

Fig. 9.19 **a** Selected frames of the swelling process of the porated cell in 35 s. The scale bar is 10 μm. **b** Variation of cell volume ratio as a function of time. Reproduced from [10] with permission from the Royal Society of Chemistry

assessment in poration studies, restricting analysis to permanent membrane damage, which is a limitation consistent with earlier laser-induced bubble approaches in microfluidics. Optimized laser cavitation parameters yield viability exceeding 90%, consistent with established findings [15]. Viability testing requires non-toxic alternatives like inert absorbing dyes, dielectric-breakdown lasers generating microbubbles via short pulses, or particle-mediated absorption. Plasmonic gold nanoparticles exemplify the latter, producing ns-pulse cavitation bubbles suitable for viability analysis [16]. Laser-induced thermal shock represents a secondary viability concern. While 80% of pulse energy heats the liquid, the thermal impact remains confined to the focal area and dissipates rapidly through adiabatic cooling during bubble expansion [17]. Parallel laser studies confirm mechanical stress as the primary effector, with thermal contributions being negligible [18].

9.4 Droplet-Based Optofluidic System for Bacteriophage Detection

Bacteriophages are viruses that infect bacterial cells. They serve as established water quality biomarkers due to their strong correlation with pathogenic virus concentrations in aquatic systems [19, 20]. Beyond environmental monitoring, bacteriophage contamination poses significant biotechnological risks, potentially invalidating microbial cultures and facilitating cross-laboratory dissemination.

The conventional plaque assay is widely used for bacteriophage quantification [21]. This method involves serial dilution of water samples, inoculation onto host-cell seeded agar plates, and incubation to allow phage propagation. The agar matrix confines progeny bacteriophages to adjacent cells, forming characteristic plaques. While providing direct visual confirmation, the technique suffers from critical constraints, including lengthy processing (24–48 h), volume limitations (1 ml sample/plate), and mandatory pre-quantification. Current adaptations combining plaque assays with PCR amplification [20, 22] retain these limitations while adding complexity through required pre-concentration steps. Emerging alternatives like electric DNA chips and functionalized quartz crystal microbalances [23, 24] offer enhanced sensitivity but suffer from narrow phage specificity and sophisticated manufacturing requirements. This creates an ongoing need for versatile, economical detection systems capable of identifying diverse, low-concentration bacteriophages.

This section presents a novel label-free bacteriophage detection platform using droplet optofluidic technology [25]. The system employs host-cell-encapsulated microdroplets as discrete bioreactors for phage infection monitoring (Fig. 9.20). The detection protocol initiates with mixing aqueous samples with *E. coli* culture medium, followed by high-throughput generation of monodisperse droplets via microfluidic fractionation. Within the microchannel network, rapid internal mixing within individual droplets promotes efficient phage-host interactions. When λ bacteriophages are present, they selectively bind to *E. coli* surface receptors and inject their genomic DNA into host cells. Subsequent intracellular replication produces progeny phages through the characteristic lytic cycle. The detection mechanism relies on optical monitoring of growth inhibition effects. Phage-containing droplets exhibit markedly reduced bacterial proliferation compared to control droplets, altering their light scattering signatures. The optofluidic platform demonstrates superior bacteriophage detection capabilities through two fundamental advantages. First, the microfluidic environment creates turbulent mixing conditions that significantly increase phage-host encounter frequency compared to static agar assays. Second, eliminating preparatory steps (e.g., serial dilution, agar plating) provides faster processing and less inaccuracies. This integration of microfluidic compartmentalization with optical detection establishes a robust framework for rapid, label-free phage quantification.

In the proposed system, the bacteriophage-containing sample combines with a host cell suspension (5×10^8 cells/mL) within the microchannel. Mineral oil subsequently emulsifies this mixture into discrete microdroplets through continuous flow generation. These droplets undergo rapid internal mixing during microchannel transit, promoting efficient phage-host interaction. Following droplet formation, the system directs the emulsion into an incubation chamber featuring mineral oil extraction ports, enabling high-density droplet storage. Post-incubation, a controlled valve mechanism releases droplets into the optical detection zone. The PDMS-based chip, fabricated via standard soft lithography, incorporates a He–Ne laser system ($\lambda = 632.8$ nm, 800 mW) with beam-splitting optics that focus incident light onto individual droplets. Scattered light signals are acquired at an 80 mm working distance using a CCD line-scan camera, with subsequent signal processing enabling quantitative analysis.

Figure 9.21 illustrates the typical scattering pattern from microdroplets containing varying concentrations of *E. coli* inclusions (rod-shaped, $d = 0.5$ μm, $l = 2$ μm). Control measurements of inclusion-free microdroplets exhibit characteristic Mie scattering fringes. In contrast, microdroplets containing bacterial inclusions demonstrate modified scattering profiles resulting from the coherent superposition of primary Mie scattering from the droplet matrix, and secondary scattering events

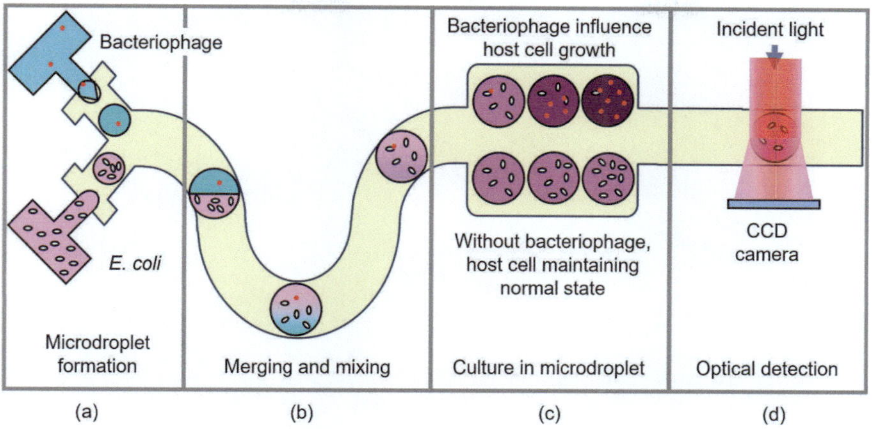

Fig. 9.20 Schematic illustration of optofluidic chip for bacteriophage detection that consists of: **a** water sample containing bacteriophage and *E. coli* in culture medium are split into microdroplets by microfluidic structure, **b** two microdroplets merge together and water sample is mixed with host cell *E. coli*, **c** host cell growth condition in the microdroplet is affected by the presence of the bacteriophage and **d** host cell growth condition is detected by using optical detection system. Reproduced from [25] with permission from the Royal Society of Chemistry

caused by light-inclusion interactions (reflection/refraction at inclusion boundaries). This combined optical effect manifests as classical Mie fringe patterns superimposed with random speckle features. Quantitative analysis reveals concentration-dependent modulation. Increasing *E. coli* loading produces progressive deviation from pristine Mie patterns, while enhanced high-frequency speckle components dominate at elevated bacterial concentrations.

Figure 9.22 demonstrates the characteristic scattering profiles generated by microdroplets (diameter: 60 μm) containing gradient concentrations of *E. coli*. The observed diffraction patterns arise from Fraunhofer diffraction at the circular droplet interface, and Mie scattering by intracellular constituents. The primary circular diffraction features are governed by droplet geometry, while intensity modulation stems from refractive index contrast between the immersion oil ($n_{\text{oil}} = 1.467$) and the aqueous phase inside droplet. This aqueous phase exhibits concentration-dependent optical properties, with its effective refractive index (n_{droplet}) increasing proportionally to bacterial load. It changes from a baseline of 1.3331 for sterile culture medium to higher values with *E. coli* colonization. Microdroplet dimensional control is achieved through precise regulating the flow rate of the two immiscible liquids. In the employed microfluidic system with fixed geometry, the generated droplet diameter remains invariant under constant flow rate conditions. This dimensional stability arises from the dominant role of channel architecture in droplet formation, exhibiting independence from variations in the viscosity ratio and interfacial tension between the immiscible fluid phases.

The coefficient of variation (CV) for droplet diameter is maintained below 3%, confirming high monodispersity in the system. Geometric tolerances for droplet

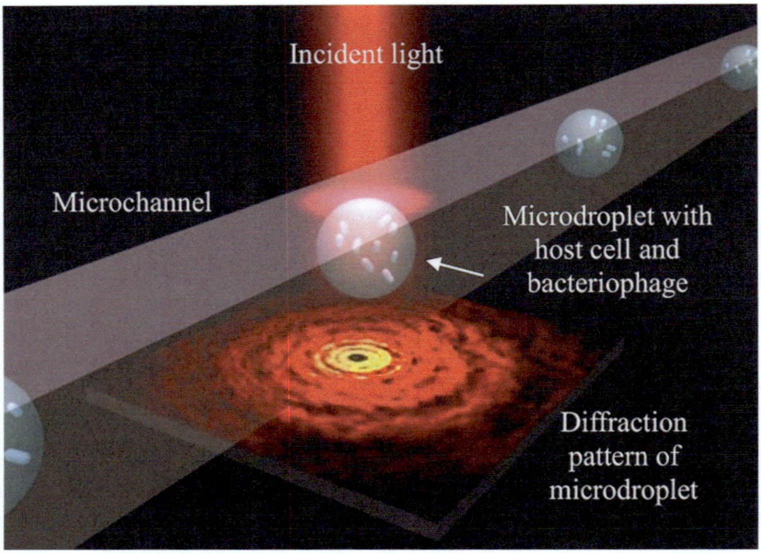

Fig. 9.21 Schematic illustration of the optofluidic imaging of droplets encapsulated with co-cultured bacteriophage and host cells, *E. coli*. Reproduced from [25] with permission from the Royal Society of Chemistry

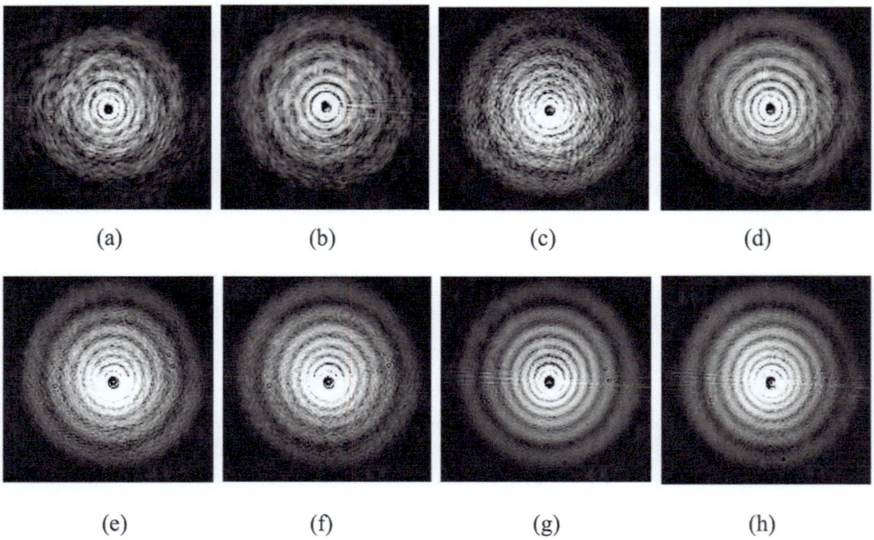

Fig. 9.22 The scattering pattern of microdroplets containing different concentration of *E. coli*: **a** 10^9 cell/ml **b** 5×10^8 cell/ml **c** 2×10^8 cell/ml **d** 10^8 cell/ml **e** 5×10^7 cell/ml **f** 2×10^7 cell/ml **g** 10^7 cell/ml and **h** control group. Reproduced from [25] with permission from the Royal Society of Chemistry

diameter, shape, and size remain invariant, allowing scattering patterns to be predominantly governed by host cell concentration. The volumetric occupancy ratio of individual *E. coli* cells within a microdroplet carrier is maintained below 0.01%, rendering the bacteriophage contribution (picoliter-scale volume) negligible in optical calculations. Even at extreme bacterial concentrations (10^9 cells/ml, ~100 cells/droplet), the effective refractive index shift is undetectable when compared to near-empty droplets (10^7 cells/ml, ≤ 1 cell/droplet).

However, cell concentration critically modulates the scattering pattern through coherent interactions of light undergoing multiple reflections and refractions in the droplet. Using a fixed-wavelength illumination source ($\lambda = 632.8$ nm) with monodisperse *E. coli* cells in microdroplets (size comparable to λ), concentration-dependent scattering patterns were recorded. The Mie scattering patterns evolves from ordered fringes to disordered speckles as cell density increases from 10^7 to 10^9 cells/ml. The transition manifests through progressive fringe extinction at high concentrations, where multiple scattering events generate high-frequency speckle noise that dominates the signal, as shown in Fig. 9.22. This phenomenon arises from constructive interference of multiply-refracted waves and phase randomization when optical depth exceeds the coherence length, effectively masking the original fringe pattern through cumulative scattering contributions from densely packed inclusions.

Consider the Fast Fourier Transform (FFT) for the frequency spectrum of the scattering patterns, which is expressed as

$$IF = \sum_{u=1}^{M} \sum_{v=1}^{M} I_s(x, y) \exp\left(-i\frac{2\pi ux}{M}\right) \exp\left(-j\frac{2\pi vy}{M}\right), \tag{9.8}$$

where I_s is the two-dimensional distribution of the scattering light. The power spectrum is

$$\varphi(f) = IF * IF, \tag{9.9}$$

where f is the frequency of intensity variation. The mean power frequency (MPF) is the normalized first order power spectral moment. It can be expressed as

$$MPF = \frac{\int_0^\infty f\varphi(f)df}{\int_0^\infty \varphi(f)df}, \tag{9.10}$$

where f is the frequency of intensity variation. MPF is used to analyze the variation of frequency of the scattering pattern.

As shown in Fig. 9.23, the mean power frequency of microdroplet scattering patterns exhibits concentration-dependent enhancement with increasing *E. coli* load (0–100 cells/droplet). This correlation enables quantitative bacterial enumeration via optical scattering analysis. A near-linear response dominates the 20–100 cells/droplet range, ensuring high measurement resolution. For the nonlinear region from 0 to 20 cells per droplet, more data are needed for refinement through additional calibration measurements. The detection resolution of this system is primarily constrained by the signal quality of captured scattering patterns. Current performance achieves single-cell sensitivity for bacterial loads of 0–20 cells/droplet, while maintaining 5% population change detection (equivalent to 2 cells at 40 cells/droplet) across the 21–100 cells/droplet range. The detection performance may be substantially enhanced through the implementation of CCD cameras featuring improved temporal resolution and photon detection efficiency.

Scattering patterns can also be classified through AI-based approaches. As shown in Fig. 9.24, principal component analysis (PCA) effectively distinguishes the scattering profiles of single microdroplets with varying *E. coli* concentrations. PCA is a statistical dimensionality reduction technique that identifies latent patterns in high-dimensional data by transforming correlated variables into a set of uncorrelated principal components (PCs), where the first few PCs capture maximal data variance [26, 27]. Widely applied in fields like image compression and facial recognition, PCA enables efficient data visualization and feature extraction. By optimizing PCA parameters, the study achieved clear separation of scattering patterns, confirming its utility for quantitative droplet analysis.

E. coli concentrations in microdroplets, including cultured alone and with λ bacteriophages, were quantified by correlating bacterial cell counts with scattering pattern signatures. Isolated *E. coli* cultures exhibited growth kinetics consistent with classical bacterial growth curves. In contrast, co-culture with λ phages in picoliter droplets

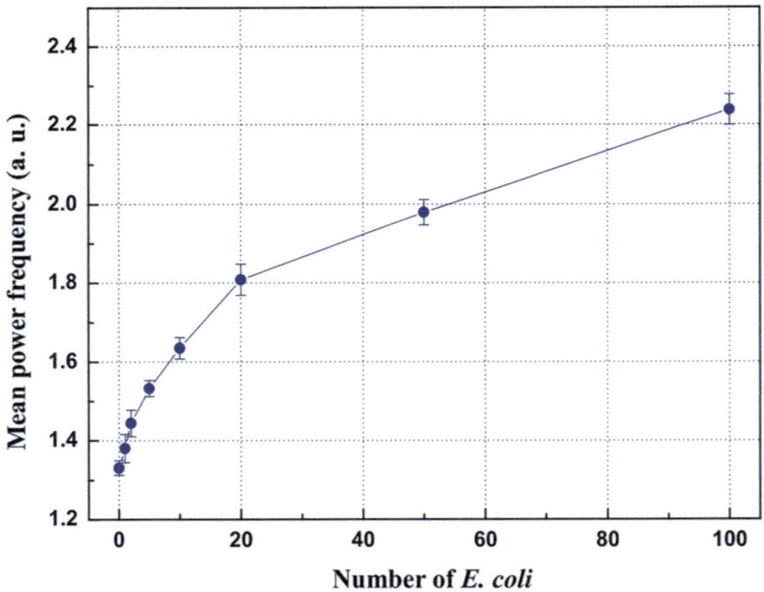

Fig. 9.23 Mean power frequency changed by the *E. coli* concentration in the microdroplet. Reproduced from [25] with permission from the Royal Society of Chemistry

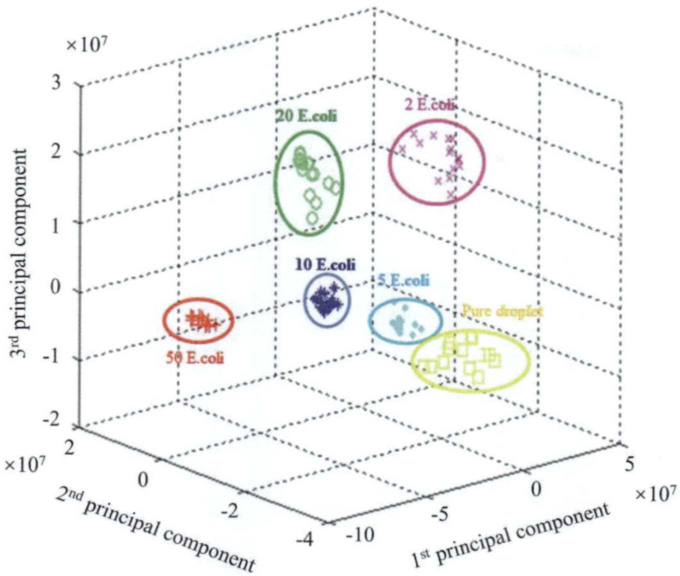

Fig. 9.24 Scattering pattern analysis with principal component analysis (PCA)

shortened the initial infection phase compared to bulk flask cultures. This acceleration stems from the confined microenvironment. The nanoscale droplet volume increases phage-host collision frequency, while microchannel flow dynamics enhance fluid mixing, collectively promoting faster infection rates.

The detection capability of the optofluidic system was validated by analyzing water samples spiked with varying bacteriophage concentrations. Phage-containing samples were injected into the system, mixing with *E. coli* culture medium (initial concentration: 5×10^8 cells/ml, ~50 cells/droplet) to create λ bacteriophage infection-permissive conditions. The mixture was divided into uniformly sized microdroplets ($d = 60\ \mu$m, $V \approx 10^{-7}$ ml), then incubated for 4 h in a microculture chamber. After that, the microdroplets were released for optical detection. Scattering pattern analysis revealed infected microdroplets (containing λ bacteriophage) exhibited reduced mean power frequency values compared to controls.

A positive detection event is registered when the specified scattering pattern is identified. The assay partitions microdroplets into multiple testing groups (N droplets per group), where the count of positive signals (f) in each group correlates with λ bacteriophage concentration ($C_{\text{bacteriophage}}$) through Eq. (9.11) [28]:

$$f = C_{\text{bacteriophage}} \times V \times N. \tag{9.11}$$

Here, V is the volume of an individual microdroplet.

Figure 9.25 presents the signal recordings from 10 microdroplet groups (1×10^5 droplets per group) tested with water samples containing λ bacteriophage at known

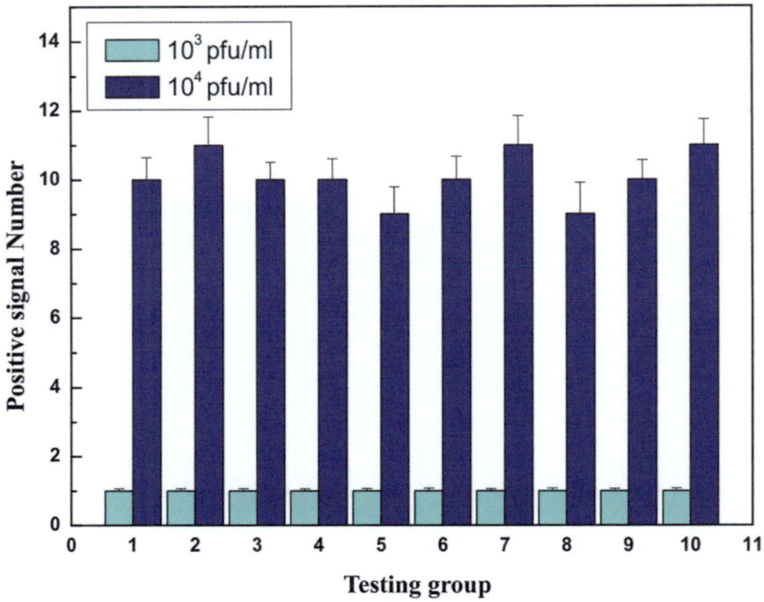

Fig. 9.25 Signal recording of λ phage containing sample with the concentration of 10^3 and 10^4 pfu/ml. Reproduced from [25] with permission from the Royal Society of Chemistry

concentrations ($C_{\text{bacteriophage}} = 10^3$ pfu/ml and 10^4 pfu/ml). The observed average positive signal counts matched theoretical predictions, with 1 and 10 signals for the 10^3 pfu/ml and 10^4 pfu/ml samples, respectively. These results confirm that the optofluidic system is capable for rapid and reliable bacteriophage detection. The detection limit can be further improved by increasing the microdroplet count per group, though its resolution is primarily constrained by the signal processing stability. Notably, this platform enables real-time, label-free quantification of bacteriophages with high sensitivity, achieving precise monitoring of host cell growth dynamics.

9.5 Summary

Optofluidic devices and applications are generally based on the integration of optical technologies (for sensing, actuation or imaging) within microfluidic (or nanofluidic) systems, allowing the use of small sample volumes in a highly controlled environment. This chapter introduces several optofluidic devices used in different applications which illustrate the diversity and multifunction of optofluidic devices. Recent years, optofluidic devices have been extensively exploited for novel functions and promising applications in chemical and biological analysis, imaging and energy. Many researchers are now trying to solve the problem of creating easy-to-use devices. Current optofluidic devices typically require extensive laboratory testing, involving additional instruments such as laser sources, thermal controllers, objectives, micropumps, and cameras. These devices also demand skilled researchers for operation. Integrating all these functionalities into a compact microsystem would significantly enhance their accessibility, making them more user-friendly and better suited for point-of-care applications.

References

1. Ghosh N, Buddhiwant P, Uppal A, Majumder SK, Patel HS, Gupta PK (2006) Simulataneous determination of size and refractive index of red blood cells by light scattering measurements. Appl Phys Lett 88:084101
2. Ito S, Takagi M (2009) Correlation between cell cycle phase of adherent Chinese hamster ovary cells and laser phase shift determined by phase-shifting laser microscopy. Biotech Lett 31:39–42
3. Yanase Y, Hiragun T, Kaneko S, Gould HJ, Greaves MW, Hide M (2010) Detection of refractive index changes in individual living cells by means of surface plasmon resonance imaging. Biosens Bioelectron 26:674–681
4. Chin LK, Liu AQ, Lim CS, Lin CL, Ayi TC, Yap PH (2010) An optofluidic volume refractometer using Fabry-Pérot resonator with tunable liquid microlenses. Biomicrofluidics 4:024107
5. Suzuki H, Sugimoto M, Matzui Y, Kondoh J (2008) Effects of gold film thickness on spectrum profile and sensitivity of a multimode-optical-fiber SPR sensor. Sens Actuators B 132:26–33
6. Curcio JA, Petty CC (1951) The near infrared absorption spectrum of liquid water. J Opt Soc Am 41:302–304
7. Cai DK, Neyer A, Kuckuk R, Heise HM (2008) Optical absorption in transparent PDMS materials applied for multimode waveguides fabrication. Opt Mater 30:1157–1161
8. Lauterborn W, Bolle H (1975) Experimental investigations of cavitation-bubble collapse in the neighbourhood of a solid boundary. J Fluid Mech 72:391–399
9. Khoo BC, Klaseboer E, Hung KC (2005) A collapsing bubble-induced micro-pump using the jetting effect. Sens Actuators A 118:152–161
10. Li ZG, Liu AQ, Klaseboer E, Zhang JB, Ohl CD (2013) Single cell membrane poration by bubble-induced microjets in a microfluidic chip. Lab Chip 13:1144–1150
11. Becker AA (1992) The boundary element method in engineering, a complete course. McGraw-Hill
12. Brebbia CA, Dominguez J (1992) Boundary elements an introductory course. WIT Press
13. Monat C, Domachul P, Eggleton BJ (2007) Integrated optofluidics: a new river of light. Nat Photonics 1:106–114
14. Salzman GC, Crowell JM, Martin JC, Trujillo TT, Romero A, Mullaney PF, LeBauve PM (1975) Cell classification by laser light scattering: identification and separation of unstained leukocytes. Acta Cytol 19:374–377
15. Galloway WJ (1954) An experimental study of acoustically induced cavitation in liquids. J Acoust Soc Am 26:849–857
16. Yuan F, Sankin G, Zhong P (2011) Dynamics of tandem bubble interaction in a microfluidic channel. J Acoust Soc Am 130:3339–3346
17. Vogel A, Noack J, Huttman G, Paltauf G (2005) Mechanisms of femtosecond laser nanosurgery of cells and tissues. Appl Phys B 81:1015–1047
18. Vogel A, Busch S, Jungnickel K, Birngruber R (1994) Mechanisms of intraocular photodisruption with picosecond and nanosecond laser pulses. Lasers Surg Med 15:32–43
19. Schijven JF, Teunis PF, Rutjes SA, Bouwknegt M, de RodaHusman AM (2011) QMRAspot: a tool for quantitative microbial risk assessment from surface water to potable water. Water Res 45(17):5564–5576

20. Ogorzaly L, Gantzer C (2006) Development of real-time RT-PCR methods for specific detection of F-specific RNA bacteriophage genogroups: application to urban raw wastewater. J Virol Methods 138(1–2):131–139
21. Loehr RC, Schwegler DT (1965) Filtration method for bacteriophage detection. Appl Microbiol 13(6):1005–1009
22. Cornax RR, Moriñigo MA, Paez IG, Muñoz MA, Borrego JJ (1990) Application of direct plaque-assay for detection and enumberation of bacteriophages of baterioides-fragilis from contaminated-water samples. Appl Environ Microbiol 56(10):3170–3173
23. Gabig-Ciminska M, Los M, Holmgren A, Albers J, Czyz A et al (2004) Detection of bacteriophage infection and prophage induction in bacterial cultures by means of electric DNA chips. Anal Biochem 324(1):84–91
24. Dultsev FN, Speight RE, Fiorini MT, Blackburn JM, Abell C, Ostanin VP, Klenerman D (2001) Direct and quantitative detection of bacteriophage by 'Hearing' surface detachment using a quartz crystal microbalance. Anal Chem 73(16):3935–3939
25. Yu JQ, Huang W, Chin LK, Lei L, Lin ZP, Ser W, Chen H, Ayi TC, Yap PH, Chen CH, Liu AQ (2014) Droplet optofluidic imaging for λ-bacteriophage detection via co-culture with host cell Escherichia coli. Lab Chip 14:3519–3524
26. Wold S, Esbensen K, Geladi P (1987) Principle component analysis. Chemom Intell Lab Syst 2(1–3):37–52
27. Price AL, Patterson NJ, Plenge RM, Weinblatt ME, Shadick NA et al (2006) Principal components analysis corrects for stratification in genome-wide association studies. Nat Genet 38(8):904–909
28. Gonzalez RC, Woods RE (2007) Digital image processing using MATLAB. Wiley-Interscience Publication

MIX
Papier aus verantwortungsvollen Quellen
Paper from responsible sources
FSC® C105338

If you have any concerns about our products,
you can contact us on
ProductSafety@springernature.com

In case Publisher is established outside the EU,
the EU authorized representative is:
Springer Nature Customer Service Center GmbH
Europaplatz 3, 69115 Heidelberg, Germany

Printed by Libri Plureos GmbH
in Hamburg, Germany